D1712483

# Transients in Power Systems

# Transients in Power Systems

**Lou van der Sluis**
*Delft University of Technology*
*The Netherlands*

JOHN WILEY & SONS, LTD
Chichester • New York • Weinheim • Brisbane • Singapore • Toronto

Baffins Lane, Chichester
West Sussex, PO19 1UD, England

National 01243 779777
International (+44) 1243 779777
e-mail (for orders and customer service enquiries): cs-books@wiley.co.uk

Visit our Home Page on http://www.wiley.co.uk or http://www.wiley.com



*Other Wiley Editorial Offices*

John Wiley & Sons, Inc., 605 Third Avenue,
New York, NY 10158-0012, USA

Wiley-VCH Verlag GmbH
Pappelallee 3, D-69469 Weinheim, Germany

John Wiley, Australia, Ltd, 33 Park Road, Milton,
Queensland 4064, Australia

John Wiley & Sons (Canada) Ltd, 22 Worcester Road
Rexdale, Ontario M9W 1L1, Canada

John Wiley & Sons (Asia) Pte Ltd, 2 Clementi Loop #02-01,
Jin Xing Distripark, Singapore 129809

***Library of Congress Cataloguing-in-Publication Data***

Van der Sluis, Lou.
Transients in power systems / by Lou van der Sluis.
p.cm.
Includes bibliographical references and index.
ISBN 0-471-48639-6
1. Transients (Electricity). 2. Electric power system stability. 3. Electric network analysis. I. Title

TK3226.V23 2001
621.31 – dc21

2001026198

***British Library Cataloguing in Publication Data***

A catalogue record for this book is available from the British Library

ISBN 0 471 48639 6

Typeset in 11/13.5pt Sabon by Laser Words, Madras, India
Printed and bound in Great Britain by Antony Rowe Ltd., Chippenham, Wiltshire
This book is printed on acid-free paper responsibly manufactured from sustainable forestry, in which at least two trees are planted for each one used for paper production.

*To*

*Linda, Tia, and Mai*

# Contents

# Preface

The power system is one of the most complex systems designed, built, and operated by engineers. In modern society, the power system plays an indispensable role, and a comparable quality of life without a constant and reliable supply of electricity is almost unthinkable. Because electricity cannot be stored in large quantities, the operation of the power system has the constraint of balancing the production of electricity in the connected power stations and the consumption by the connected loads and of maintaining constant frequency and constant voltage with the clients. During normal operation, loads are connected and disconnected. Control actions are therefore continuously necessary – the power system is never in a steady state. On a timescale of years, planning of new power plants, the erection of new transmission lines, or the upgrading from existing lines to higher voltage levels are important items to consider. When we look ahead into the future, the main topic is the economical operation – what is the expected load and what is the most economical fuel to be used to heat the boilers in the power stations. When the reliability of the system is analysed with repetitive load-flow calculations, the timescale is usually hours, yet when the dynamic stability is analysed to verify whether the system remains stable after a major disturbance, the power system is studied with an accuracy of seconds. Switching actions, either to connect or disconnect loads or to switch off faulted sections after a short-circuit, and disturbances from outside, such as a lightning stroke on or in the vicinity of a high-voltage transmission line, make it necessary to examine the power system on an even smaller timescale, microseconds to milliseconds. We speak in that case of *electrical transients*. The time that the electrical transients are present in the system is short, but during a transient period, the components in the system are subjected to high current and high-voltage peaks that can cause considerable damage.

This book deals with electrical transients in the power system. Much has been learned about transient phenomena since the early days of power system operation. Pioneers in this field were men like Charles Proteus Steinmetz and Oliver Heaviside who focussed on the understanding of electrical transients in a more or less general way. They took the analytical approach, which is restricted to linear circuits. When a circuit becomes more complex, the application of this method becomes very laborious and time-consuming. After the Second World War, new tools were developed and used in studying circuit transient phenomena that were previously avoided because of their complexity. The transient network analyser (TNA) was exceptionally useful in studying the behaviour of a large variety of complex linear and nonlinear circuits. The TNA was a powerful tool for obtaining solutions to problems involving distributed constants as well as nonlinear impedances. The use of the analogue TNA resulted in the publication of much technical literature. In 1951, Harold Peterson published his book '*Transients in Power Systems*' with many examples of TNA studies. Peterson's book is a practical survey of the particular phenomena (faults, sudden loss of load, switching surges, and so forth.) that can cause transients and is based on his practical experience with the General Electric Company in the USA. A classical book is Reinhold Ruedenberg's '*Transient Performance of Power Systems*,' published in 1950 and based on his earlier work written in German. In addition, switchgear design is closely related to electrical transient phenomena, and books from authors such as Biermanns and Slamecka, who wrote from their experience with the switchgear divisions of AEG and Siemens, are a historical source for the understanding of transient phenomena and switchgear development.

When I joined KEMA in 1977, as a test engineer in the famous 'de Zoeten' high-power laboratory, I entered the world of short-circuit testing. The testing of power system equipment according to IEC and ANSI standards, calculating test circuits, measuring high currents and high voltages in an electromagnetically hostile environment, and so forth deepened my knowledge about electrical engineering and about physics. My first introduction to the subject was Allan Greenwood's '*Electrical Transients in Power Systems*.' Later, I went through many more classical books and papers, which gave me a good overview of the historical development of high-voltage circuit breakers. In the fifteen years and more that I had the pleasure of working at KEMA, I learned a lot from my former colleagues at the high-power laboratory. Together we designed new test circuits, developed new measuring equipment and built a computerised measurement system with transient recorders and computer workstations.

KEMA's high-power laboratory has always been a front-runner when it comes to test circuit development and there has always been a strong participation in IEC standardisation work.

In 1990, I joined the Delft University of Technology as a part-time professor to teach a course in transients in power systems. Since 1992, I have been a full-time professor and head of the Power Systems Laboratory. I am always pleased to know that students are very much interested in switching phenomena and attracted by the operation of high-voltage circuit breakers and the physical processes that take place during current interruption at current zero. The advanced mathematics, together with physics and sometimes exploding equipment is probably the right mixture. After a couple of years, lecture notes need an update. In 1996, Adriaan de Lange contacted me about a PhD research project, and to refresh his knowledge about switching transients, he attended my lectures and made notes of what I explained to the audience that was not written down in the lecture notes. Adriaan also researched extensively literature on circuit breaker development, current zero phenomena, and testing techniques to acquire a solid base for his thesis. Without the effort of Adriaan, this book would not have been written.

Chapter 1, *Basic Concepts and Simple Switching Transients*, summarises the fundamental physical phenomena and the mathematical tools to tackle transient phenomena. In fact, basic network theory and a thorough understanding of simple LR and RLC networks and the behaviour of the transient voltages and currents after a switching action is a necessity. When analysing transients, one always tries to reduce complex networks to series or parallel networks for a first-approximation. The three-phase layout of the power system is treated in Chapter 2, *Transient Analysis of Three-Phase Power Systems*, wherein faults that result in severe system stresses are analysed with symmetrical component networks. The properties of travelling waves, which play an important role in the subject, are treated in Chapter 3, *Travelling Waves*. Overvoltages caused by operation of high-voltage circuit breakers can only be predicted when the physical processes between the breaker contacts and the influence of the different extinguishing media on the current interruption is understood. The different high-voltage circuit breakers, the current interruption process, and arc–circuit interaction are described in Chapter 4, *Circuit Breakers*. In Chapter 5, *Switching Transients*, the current and voltage oscillations that occur most often in practice, such as capacitive current interruption, capacitive inrush currents, the interruption of small inductive currents, transformer inrush currents, and the short-line fault are treated. Chapter 6, *Power System Transient Recovery Voltages* describes power frequency transients that quite often result from switching actions and that can cause considerable damage to the power system components. An overview of how the different short-circuit duties are represented in the IEC and IEEE/ANSI standards is given. In Chapter 7, *Lightning-Induced Transients*, the mechanism of lightning is explained, and the chapter focuses on the impact of lightning strokes on or in the vicinity of transmission lines and substations. The calculation of electrical transients without the help of a computer is nowadays hardly unthinkable. The mathematical formulation and the numerical treatment of power system transients is shown in Chapter 8, *Numerical Simulation of Electrical Transients.* Special attention is given on how to incorporate nonlinear elements, such as arc models, in transient computer programs such as EMTP, MNA, and XTrans and the MATLAB Power System Blockset. A demo version of the XTrans program can be downloaded from *http://eps.et.tudelft.nl.* The background of the insulation coordination and the relevant IEC-standards and IEEE/ANSI-standards together with a brief history of IEC, ANSI, CIGRE and STL are given in Chapter 9, *Insulation Coordination, Standardisation Bodies, and Standards.* Testing of high-voltage circuit breakers (the proof of the pudding is in the eating) in the high-power laboratory and the related measurements and measuring equipment are described in Chapter 10, *The Testing of Circuit Breakers.*

I am very much obliged to my secretary Tirza Drisi who devotedly edited the manuscript and to Henk Paling who made the excellent drawings. Pieter Schavemaker put many hours in painstakingly reading the manuscript to filter out errors and inconsistencies. In Chapter 8, *The Numerical Simulation of Electrical Transients*, I fruitfully used the educational and illustrative examples that Pieter developed for the chapter about numerical transient calculations in his thesis. In writing Chapter 6, *Power System Transient Recovery Voltages*, I received valuable support about the latest developments in IEC and IEEE/ANSI standards from Henk te Paske, Test Engineer at KEMA's high-power laboratory. Martijn Venema from KEMA supplied the photos.

**Lou van der Sluis**
*Nootdorp, Spring 2001*

# 1

# Basic Concepts and Simple Switching Transients

The purpose of a power system is to transport and distribute the electrical energy generated in the power plants to the consumers in a safe and reliable way. Aluminium and copper conductors are used to carry the current, transformers are used to bring the electrical energy to the appropriate voltage level, and generators are used to take care of the conversion of mechanical energy into electrical energy. When we speak of *electricity*, we think of current flowing through the conductors from generator to load. This approach is valid because the physical dimensions of the power system are large compared with the wavelength of the currents and voltages; for 50-Hz signals, the wavelength is 6000 km. This enables us to apply Kirchhoff's voltage and current laws and use lumped elements in our modelling of the power system. In fact, the transportation of the electrical energy is done by the electromagnetic fields that surround the conductors and the direction of the energy flow is given by the Poynting vector.

For steady-state analysis of the power flow, when the power frequency is a constant 50 or 60 Hz, we can successfully make use of complex calculus and phasors to represent voltages and currents. Power system transients involve much higher frequencies up to kiloHertz and megaHertz. Frequencies change rapidly, and the complex calculus and the phasor representation cannot be applied any longer. Now the differential equations describing the system phenomena have to be solved. In addition, the lumped-element modelling of the system components has to be done with care if we want to make use of Kirchhoff's voltage and current laws. In the case of a power transformer, under normal power frequency–operation conditions, the transformer ratio is given by the ratio between the number of the windings of the primary coil and the number of the windings of the secondary coil. However, for a lightning-induced voltage wave, the stray capacitance of the windings and the stray capacitance between the primary and secondary coil determine the transformer ratio. In these two situations, the power transformer has to be modelled differently!

When we cannot get away with a lumped-element representation, wherein the inductance represents the magnetic field and the capacitance represents the electric field and the resistance losses, we have to do the analysis by using travelling waves. The correct 'translation' of the physical power system and its components into suitable models for the analysis and calculation of power system transients requires insight into the basic physical phenomena. Therefore, it requires careful consideration and is not easy.

A transient occurs in the power system when the network changes from one steady state into another. This can be, for instance, the case when lightning hits the ground in the vicinity of a high-voltage transmission line or when lightning strikes a substation directly. The majority of power system transients is, however, the result of a switching action. Load-break switches and disconnectors switch off and switch on parts of the network under load and no-load conditions. Fuses and circuit breakers interrupt higher currents and clear short-circuit currents flowing in faulted parts of the system. The time period when transient voltage and current oscillations occur is in the range of microseconds to milliseconds. On this timescale, the presence of a short-circuit current during a system fault can be regarded as a steady-state situation, wherein the energy is mainly in the magnetic field, and when the fault current has been interrupted, the system is transferred into another steady-state situation, wherein the energy is predominantly in the electric field. The energy exchange from the magnetic field to the electric field is when the system is visualised by lumped elements, noticed by transient current and voltage oscillations.

In this chapter, a few simple switching transients are thoroughly analysed to acquire a good understanding of the physical processes that play a key role in the transient time period of a power system. As switching devices, we make use of the *ideal switch*. The ideal switch in closed position is an ideal conductor (zero resistance) and in open position is an ideal isolator (infinite resistance). The ideal switch changes from close to open position instantaneously, and the sinusoidal current is always interrupted at current zero.

## 1.1 SWITCHING AN LR CIRCUIT

A sinusoidal voltage is switched on to a series connection of an inductance and a resistance (Figure 1.1). This is in fact the most simple single-phase representation of a high-voltage circuit breaker closing into a short-circuited transmission line or a short-circuited underground cable. The voltage source $E$ represents the electromotive forces from the connected synchronous generators. The inductance $L$ comprises the synchronous inductance of these generators, the leakage inductance of the power transformers, and the inductance of the bus bars, cables, and transmission lines. The resistive losses of the supply circuit are represented by the resistance $R$. Because we have linear network elements only, the current flowing in the circuit after closing the switch can be seen as the superposition of a transient current and a steady-state current.

The transient current component is determined by the inductance and the resistance only and is not influenced by the sources in the network (in this case by the voltage source $E$). It forms the *general* solution of the first-order *homogeneous differential equation*, whereas the steady-state current component is the *particular* solution of the *nonhomogeneous differential equation*. In the latter case, the transient oscillations are damped out because their energy is dissipated in the resistive part of the circuit. Applying Kirchhoff's voltage law gives us the nonhomogeneous differential equation of the circuit in Figure 1.1:

$$E_{\max} \sin(\omega t + \varphi) = Ri + L\frac{\mathrm{d}i}{\mathrm{d}t} \tag{1.1}$$

The switch can close the circuit at any time instant and the phase angle can have a value between 0 and $2\pi$ rad. To find the general solution of the differential equation, we have to solve the *characteristic* equation of

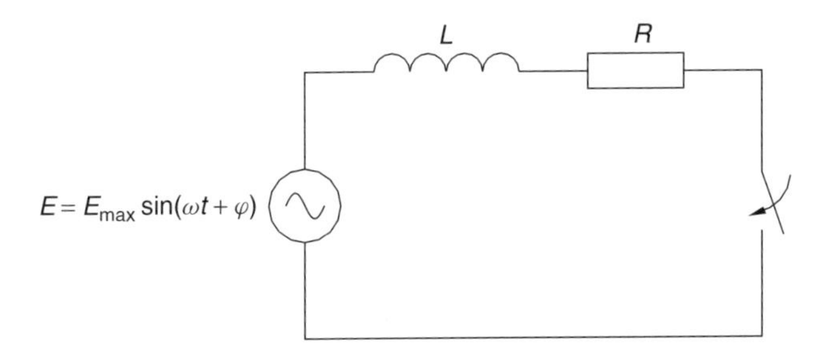


**Figure 1.1** A sinusoidal voltage source is switched on to an LR series circuit

the homogeneous differential equation

$$Ri + L\lambda i = 0 \tag{1.2}$$

The scalar $\lambda$ is the eigenvalue of the characteristic equation. We find for $\lambda = -(R/L)$, and thus the general solution for Equation (1.1) is

$$i_h(t) = C_1 e^{-(R/L)t} \tag{1.3}$$

The particular solution is found by substituting in Equation (1.1) a general expression for the current

$$i_p(t) = A\sin(\omega t + \varphi) + B\cos(\omega t + \varphi) \tag{1.4}$$

$A$ and $B$ can be determined:

$$A = \frac{RE_{\max}}{R^2 + \omega^2 L^2} \quad B = -\frac{\omega L E_{\max}}{R^2 + \omega^2 L^2} \tag{1.5}$$

This results in the particular solution for the current

$$i_p(t) = \frac{E_{\max}}{\sqrt{R^2 + \omega^2 L^2}} \sin\left[\omega t + \varphi - \tan^{-1}\left(\frac{\omega L}{R}\right)\right] \tag{1.6}$$

The complete solution, which is the sum of the general and particular solution, is

$$\begin{aligned} i(t) &= i_h(t) + i_p(t) \\ &= C_1 e^{-(R/L)t} + \frac{E_{\max}}{\sqrt{R^2 + \omega^2 L^2}} \sin\left[\omega t + \varphi - \tan^{-1}\left(\frac{\omega L}{R}\right)\right] \end{aligned} \tag{1.7}$$

Before the switch closes (Figure 1.1), the magnetic flux in the inductance $L$ is equal to zero; this remains so immediately after the instant of closing, owing to the law of the conservation of flux. Therefore, at $t = 0$, the instant of closing, we can write

$$C_1 + \frac{E_{\max}}{\sqrt{R^2 + \omega^2 L^2}} \sin\left[\varphi - \tan^{-1}\left(\frac{\omega L}{R}\right)\right] = 0 \tag{1.8}$$

This gives us the value for $C_1$; hence, the complete expression for the current becomes

$$i(t) = e^{-(R/L)t}\left\{\frac{-E_{\max}}{\sqrt{R^2+\omega^2L^2}}\sin\left[\varphi - \tan^{-1}\left(\frac{\omega L}{R}\right)\right]\right\}$$
$$+\frac{E_{\max}}{\sqrt{R^2+\omega^2L^2}}\sin\left[\omega t+\varphi - \tan^{-1}\left(\frac{\omega L}{R}\right)\right] \qquad (1.9)$$

The first part of Equation (1.9) contains the term $\exp[-(R/L)t]$ and damps out. This is called the *DC component*. The expression between the brackets is a constant and its value is determined by the instant of closing of the circuit. For $[\varphi - \tan^{-1}(\omega L/R)] = 0$ or an integer times $\pi$, the DC component is zero, and the current is immediately in the steady state. In other words, there is no transient oscillation. When the switch closes the circuit 90° earlier or later, the transient current will reach a maximum amplitude, as can be seen in Figure 1.2.

The current in Figure 1.2 is called an *asymmetrical current*. In the case where no transient oscillation occurs and the current is immediately in the steady state, we speak of a *symmetrical current*. The asymmetrical current can reach a peak value of nearly twice that of the symmetrical current, depending on the time constant $L/R$ of the supply circuit. This implies that, for instance, when a circuit breaker closes on a short-circuited high-voltage circuit, strong dynamic forces will act on the connected bus bars and lines because of the large current involved.

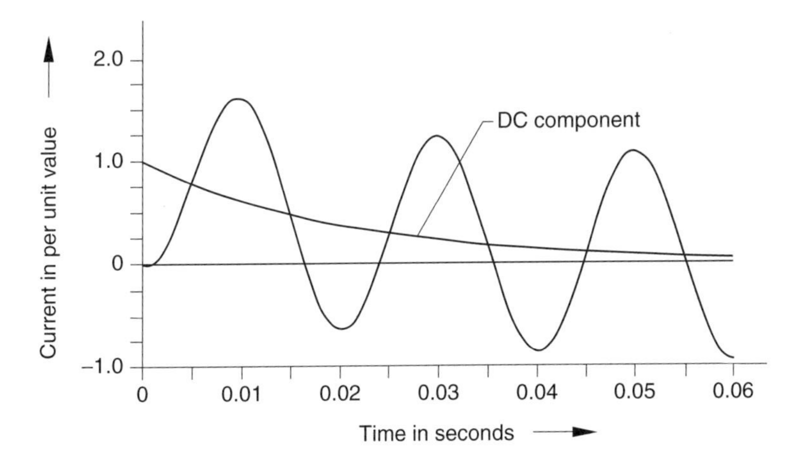


**Figure 1.2** The shape of a transient current in an inductive circuit depends on the instant of switching

When the time constant of the supply circuit is rather high, which is the case for short-circuit faults close to the generator terminals, the *transient* and *subtransient* reactance of the synchronous generator cause an extra-high first peak of the short-circuit current. After approximately 20 milliseconds, when the influence of the transient and subtransient reactance is not present any longer, the synchronous reactance reduces the root-mean-square value (rms value) of the short-circuit current. Under these circumstances, an alternating current flows without current zeros for several periods in the case of a fault in one of the phases because of the large DC component. This current cannot be interrupted because the current zero necessary for current interruption is lacking.

## 1.2 SWITCHING AN LC CIRCUIT

Another basic network is the series connection of an inductance and a capacitance; this is in fact the most simple representation of a high-voltage circuit breaker switching a capacitor bank or a cable network. To make it simple, we first analyse the case in which a DC source energises the network by closing the (ideal) switch.

As can be seen from Figure 1.3, there are two energy-storage components – the inductance storing the magnetic energy and the capacitance storing the electric energy. After closing the switch, an oscillation can occur in the network. This is due to the fact that an exchange of energy takes place between the two energy-storage devices with a certain frequency. Applying Kirchhoff's voltage law results in

$$E = L\frac{\mathrm{d}i}{\mathrm{d}t} + \frac{1}{C}\int i\,\mathrm{d}t \qquad (1.10)$$

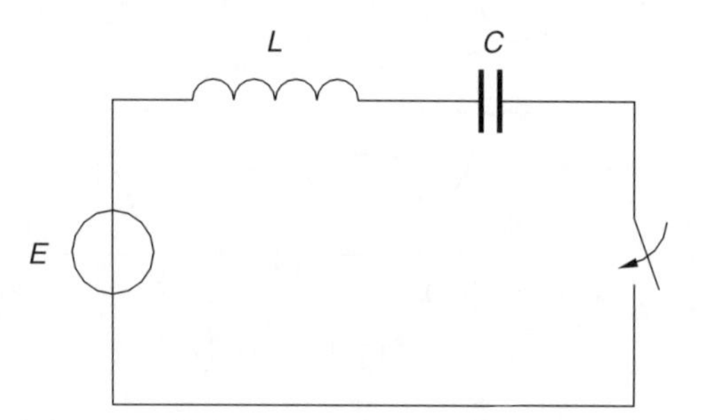


**Figure 1.3** A DC source switched on an LC series network

To solve this differential equation, it is transformed to the Laplace domain, and we get the following algebraic equation:

$$\frac{E}{p} = pLi(p) - Li(0) + \frac{i(p)}{pC} + \frac{V_c(0)}{p} \tag{1.11}$$

In this equation, $p$ is the complex Laplace variable.

This can be written as

$$i(p)\left(p^2 + \frac{1}{LC}\right) = \frac{E - V_c(0)}{L} + pi(0) \tag{1.12}$$

When we look for the initial conditions, it is clear that $i(0) = 0$ as the current in the network is zero before the switch closes and because of the physical law of conservation of the flux. This is the case immediately after closing of the switch too. In the case of the capacitor, the situation is not so easy because the capacitor can have an initial voltage, for instance, because of a trapped charge on a capacitor bank.

Let us assume that there is no charge on the capacitor and therefore $V_c(0) = 0$ and let $\omega_0^2 = 1/LC$.

Equation (1.12) becomes

$$i(p) = E\sqrt{\frac{C}{L}}\frac{\omega_0}{p^2 + \omega_0^2} \tag{1.13}$$

and back-transformation from the Laplace domain to the time domain gives the solution of Equation (1.10)

$$i(t) = E\sqrt{\frac{C}{L}}\sin(\omega_0 t) \tag{1.14}$$

In Equation (1.14), we can recognise two important properties of the LC series network.

- After closing the switch at time $t = 0$, an oscillating current starts to flow with a natural frequency

$$\omega_0 = \sqrt{LC}$$

- The *characteristic impedance*, $Z_0 = (L/C)^{1/2}$, together with the value of the source voltage $E$, determines the peak value of the oscillating current.

When there is a charge present on the capacitor, the current in the Laplace domain becomes

$$i(p) = [E - V_c(0)]\sqrt{\frac{C}{L}}\frac{\omega_0}{p^2 + \omega_0^2} \tag{1.15}$$

For the capacitor voltage in the Laplace domain

$$V_c(p) = \frac{E}{p} - pLi(p) = \frac{E}{p} - [E - V_c(0)]\frac{p}{p^2 + \omega_0^2} \tag{1.16}$$

and after back-transformation into the time domain

$$V_c(t) = E - [E - V_c(0)]\cos(\omega_0 t) \tag{1.17}$$

Figure 1.4 shows the voltage waveforms for three initial values of the capacitor voltage. From these voltage waveforms in Figure 1.4, it can be seen that for $V_c(0) = 0$ the voltage waveform has what is called a (1-cosine) shape and that it can reach twice the value of the peak of the source voltage. For a negative charge, the peak voltage exceeds this value, as the electric charge cannot change instantly after closing the switch. In addition, when the characteristic impedance of the circuit has a low value, for example, in the case of switching a capacitor bank (a large $C$) and a strong supply (a small $L$), the peak of the inrush current after closing the switch can reach a high value.

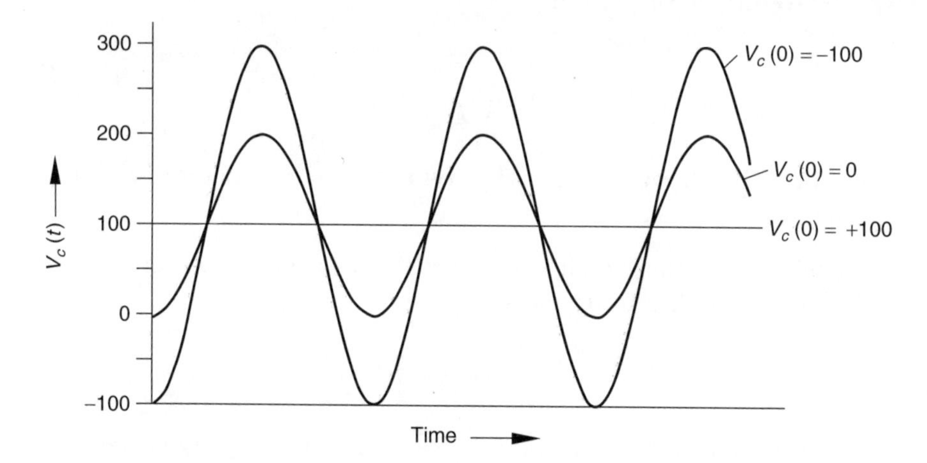


**Figure 1.4** Voltage across the capacitor for three different initial values of the capacitor voltage. The DC voltage source has the value $E = 100$ V

## 1.3 SWITCHING AN RLC CIRCUIT

In practice, there is always damping in the series circuit and that can be represented by adding a resistance in series. When a sinusoidal voltage source $E_{max}\sin(\omega t+\varphi)$ is switched on in the RLC series circuit at $t=0$ (see Figure 1.5), Kirchhoff's voltage law leads to

$$E_{max}\sin(\omega t+\varphi)=L\frac{\mathrm{d}i}{\mathrm{d}t}+Ri+\frac{1}{C}\int i\,\mathrm{d}t \tag{1.18}$$

To find the transient (or natural) responses of the network, we have to solve the homogeneous differential equation

$$0=\frac{\mathrm{d}^2 i}{\mathrm{d}t^2}+\frac{R}{L}\frac{\mathrm{d}i}{\mathrm{d}t}+\frac{1}{LC}i \tag{1.19}$$

The general solution of the homogeneous differential equation is

$$i_h(t)=C_1\mathrm{e}^{\lambda_1 t}+C_2\mathrm{e}^{\lambda_2 t} \tag{1.20}$$

where $\lambda_1$ and $\lambda_2$ are the roots of the characteristic equation

$$0=\lambda^2+\frac{R}{L}\lambda+\frac{1}{LC} \tag{1.21}$$

$$\lambda_{1,2}=-\frac{R}{2L}\pm\sqrt{\left(\frac{R}{2L}\right)^2-\frac{1}{LC}} \tag{1.22}$$

The values of the inductance, capacitance, and resistance are indeed positive because they are physical components. The absolute value of the expression $[(R/2L)^2-(1/LC)]^{1/2}$ is smaller than $R/2L$. When $(R/2L)^2-(1/LC)$ is positive, the roots $\lambda_1$ and $\lambda_2$ are negative. When $(R/2L)^2-(1/LC)$ is negative, the roots $\lambda_1$ and $\lambda_2$ are complex but the real part is

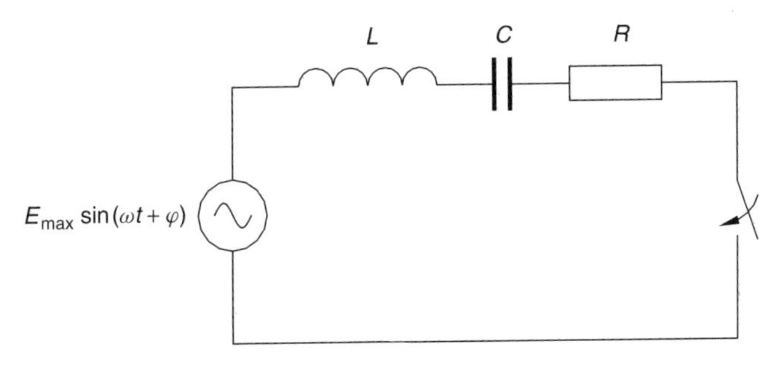


**Figure 1.5** A sinusoidal voltage source is switched on an RLC series circuit

negative. This shows that in the general solution

$$i_h(t) = C_1 e^{\lambda_1 t} + C_2 e^{\lambda_2 t} \tag{1.23}$$

the exponential functions will become zero for large values of $t$ and the *particular* solution will remain. This particular solution can be written as

$$i_p(t) = A \sin(\omega t + \varphi) + B \cos(\omega t + \varphi) \tag{1.24}$$

In this particular solution, the constants $A$ and $B$ have to be determined. Equation (1.24) is substituted in Equation (1.18) and this gives us for the particular solution

$$i_p(t) = \frac{E_{\max}}{\sqrt{R^2 + \left(\frac{1}{\omega C} - \omega L\right)^2}} \sin\left[\omega t + \varphi + \tan^{-1}\left(\frac{\frac{1}{\omega C} - \omega L}{R}\right)\right] \tag{1.25}$$

The complete solution, which is the sum of the general and particular solution, is

$$i(t) = i_h(t) + i_p(t) = (C_1 e^{\lambda_1 t} + C_2 e^{\lambda_2 t}) + \frac{E_{\max}}{\sqrt{R^2 + \left(\frac{1}{\omega C} - \omega L\right)^2}} \times \sin\left[\omega t + \varphi + \tan^{-1}\left(\frac{\frac{1}{\omega C} - \omega L}{R}\right)\right] \tag{1.26}$$

Three different situations can be distinguished:

1. When $(R/2L)^2 > 1/LC$, the transient oscillation is *overdamped* and the roots of the characteristic Equation (1.21) are both real and negative. The expression for the current becomes

$$i(t) = e^{\alpha t}(C_1 e^{\beta t} + C_2 e^{-\beta t}) + \frac{E_{\max}}{\sqrt{R^2 + \left(\frac{1}{\omega C} - \omega L\right)^2}} \sin\left[\omega t + \varphi + \tan^{-1}\left(\frac{\frac{1}{\omega C} - \omega L}{R}\right)\right] \tag{1.27}$$

with $\alpha = -(R/2L)$ and $\beta = [(R/2L)^2 - (1/LC)]^{1/2}$

2. When $(R/2L)^2 = 1/LC$, the roots of the characteristic equation are equal and real and the transient oscillation is said to be *critically damped.* The expression for the critically damped current is

$$i(t) = \mathrm{e}^{\alpha t}(C_1 + C_2) + \frac{E_{\max}}{\sqrt{R^2 + \left(\dfrac{1}{\omega C} - \omega L\right)^2}} \times \sin\left[\omega t + \varphi + \tan^{-1}\left(\frac{\dfrac{1}{\omega C} - \omega L}{R}\right)\right] \tag{1.28}$$

with $\alpha = -(R/2L)$.

3. In the case that $(R/2L)^2 < 1/LC$, the roots $\lambda_1$ and $\lambda_2$ in the general solution (Equation (1.23)) are complex.

$\lambda_1 = \alpha + j\beta$ and $\lambda_2 = \alpha - j\beta$ with $\alpha = -(R/2L)$ and $\beta = [(1/LC) - (R/2L)^2]^{1/2}$ and Equation (1.23) can be written as

$$i_h(t) = C_1\mathrm{e}^{\alpha t + j\beta t} + C_2\mathrm{e}^{\alpha t - j\beta t} \tag{1.29}$$

and because $C_2 = C_1^*$

$$i_h(t) = C_1\mathrm{e}^{\alpha t + j\beta t} + (C_1\mathrm{e}^{\alpha t + j\beta t})^* \tag{1.30}$$

Making use of the property of complex numbers that $Z + Z^* = 2\mathrm{Re}(Z)$ and using the Euler notation $\mathrm{e}^{j\beta t} = \cos(\beta t) + j\sin(\beta t)$, Equation (1.30) can be written as

$$i_h(t) = 2\mathrm{e}^{\alpha t}\mathrm{Re}(C_1\mathrm{e}^{j\beta t}) \tag{1.31}$$

with

$$C_1 = \mathrm{Re}(C_1) + j\,\mathrm{Im}(C_1)$$

Equation (1.31) can be written as

$$i_h(t) = 2\mathrm{e}^{\alpha t}\mathrm{Re}(C_1)\cos(\beta t) - 2\mathrm{e}^{\alpha t}\,\mathrm{Im}(C_1)\sin(\beta t) \tag{1.32}$$

When we substitute for $\mathrm{Re}(C_1) = (k_1/2)$ and for $\mathrm{Im}\,(C_1) = (-k_2/2)$, the expression for the general solution is $i_h(t) = \mathrm{e}^{\alpha t}(k_1\cos\beta t + k_2\sin\beta t)$. The complete solution for the oscillating current is

$$i(t) = e^{\alpha t}[k_1 \cos(\beta t) + k_2 \sin(\beta t)] + \frac{E_{max}}{\sqrt{R^2 + \left(\frac{1}{\omega C} - \omega L\right)^2}} \times \sin\left[\omega t + \varphi + \tan^{-1}\left(\frac{\frac{1}{\omega C} - \omega L}{R}\right)\right] \tag{1.33}$$

with $\alpha = -(R/2L)$ and $\beta = [(1/LC) - (R/2L)^2]^{1/2}$.

In the three cases mentioned herewith (see Figure 1.6), the particular solution is the same but the general solution is different. The transient component in the current contains sinusoidal functions with angular frequency $\beta$, which usually differs from 50- or 60-Hz power frequency of the particular solution, and this is the cause for the irregular shape of the current. When the DC component $\exp(\alpha t) = \exp[-(R/2L)t] = \exp[-(t/\tau)]$ has damped out with the damping time constant and the transient part of the current has been reduced to zero, the steady-state

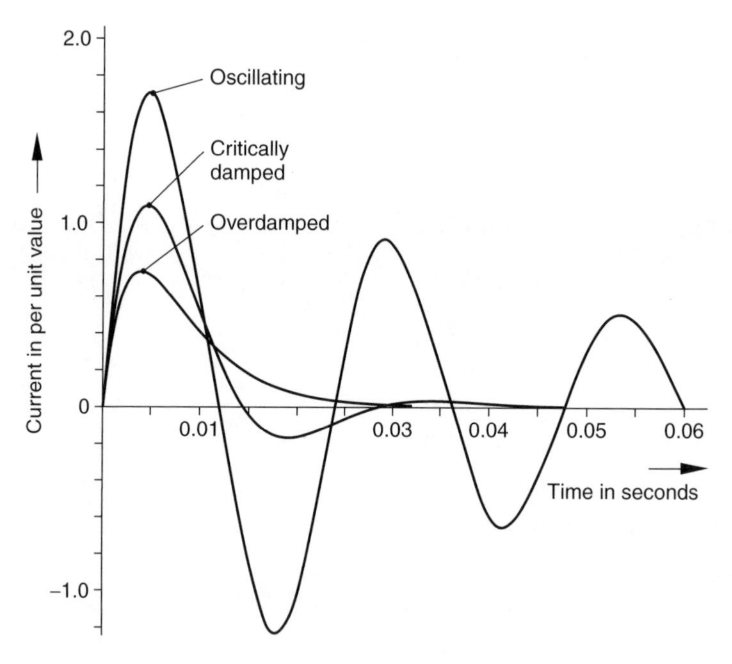


**Figure 1.6** Overdamped, critically damped, and oscillating response of an RLC series circuit after closing the switch at maximum supply voltage $t = 0$, $\varphi = +\frac{\pi}{2}$

current *lags* or *leads* the voltage of the source. The absolute value of $1/\omega C$ and $\omega L$ in the term $\tan^{-1}\{[(1/\omega C) - (\omega L)]/R\}$ determine if the current is lagging (in a dominant inductive circuit) or leading (in a dominant capacitive circuit). After a time span of three times the damping time constant $\tau = 2L/R$, only 5 percent of the initial amplitude of the transient waveform is present in the network.

It is not necessary that after every change of state, such as after a switching action, transient oscillations occur in a network. It is very well possible that the initial conditions and the instant of switching are such that immediately after closing of the switch the steady-state situation is present. A good example is the switching of a lossless reactor when the supply voltage is at the maximum. In practice, however, this rarely happens, and after switching actions, transient oscillations originate in an electrical network. Power systems have a high quality factor, that is, a large $L$ and small $R$ and are designed such that the frequency of the transient oscillations is much larger than the power frequency; this avoids steady-state overvoltages because of resonance. However, for higher harmonic frequencies generated by power electronic equipment, resonance can occur more easily.

## 1.4 REFERENCES FOR FURTHER READING

Boyce, W. E. and DiPrima, R. C., *Elementary Differential Equations and Boundary Value Problems*, 3rd ed., Chapter 3, Wiley & Sons, New York, 1977.

Edminister, J. A., *Electric Circuits*, 2nd ed., Chapter 5, McGraw-Hill, NewYork, 1983.

Greenwood, A., *Electrical Transients in Power Systems*, 2nd ed., Chapters 1–4, Wiley & Sons, New York, 1991.

Happoldt, H. and Oeding, D., *Elektrische Kraftwerke und Netze*, 5th ed., Chapter 17, Springer-Verlag, Berlin, 1978.

Ruedenberg, R., *Transient Performance of Electric Power Systems: Phenomena in Lumped Networks*, Chapter 3, McGraw-Hill, New York, 1950.

Ruedenberg, R., in H. Dorsch and P. Jacottet, eds., *Elektrische Schaltvorgaenge*, 5th ed., Vol. I, Chapters 2, 3, Springer-Verlag, Berlin, 1974.

Slamecka, E. and Waterschek, W., *Schaltvorgaenge in Hoch- und Niederspannungsnetzen*, Chapters 3, 5, Siemens Aktiengesellschaft, Berlin, 1972.

Wylie, C. R. and Barrett, L. C., *Advanced Engineering Mathematics*, 5th ed., Chapter 6, McGraw-Hill, New York, 1989.

# 2

# Transient Analysis of Three-Phase Power Systems

In normal operating conditions, a three-phase power system can be treated as a single-phase system when the loads, voltages, and currents are balanced. If we postulate plane-wave propagation along the conductors (it is, however, known from the Maxwell equations that in the presence of losses this is not strictly true), a network representation with lumped elements can be made when the physical dimensions of the power system, or a part of it, are small as compared with the wavelength of the voltage and current signals. When this is the case, one can successfully use a single-line lumped-element representation of the three-phase power system for calculation. A fault brings the system to an abnormal condition. Short-circuit faults are especially of concern because they result in a switching action, which often results in transient overvoltages.

Line-to-ground faults are faults in which an overhead transmission line touches the ground because of wind, ice loading, or a falling tree limb. A majority of transmission-line faults are single line-to-ground faults. Line-to-line faults are usually the result of galloping lines because of high winds or because of a line breaking and falling on a line below. Double line-to-ground faults result from causes similar to that of the single line-to-ground faults but are very rare. Three-phase faults, when all three lines touch each other or fall to ground, occur in only a small percentage of the cases but are very severe faults for the system and its components.

In the case of a symmetrical three-phase fault in a symmetrical system, one can still use a single-phase representation for the short-circuit and transient analysis. However, for the majority of the fault situations, the power system has become unsymmetrical. *Symmetrical components* and, especially, the *sequence networks* are an elegant way to analyse faults in unsymmetrical three-phase power systems because in many cases the unbalanced portion of the physical system can be isolated for a study, the rest of the system being considered to be in balance. This is, for instance, the case for an unbalanced load or fault. In such cases, we attempt to find the symmetrical components of the voltages and the currents at the point of unbalance and connect the sequence networks, which are, in fact, copies of the balanced system at the point of unbalance (the fault point).

## 2.1 SYMMETRICAL COMPONENTS IN THREE-PHASE SYSTEMS

In 1918, C. L. Fortescue published a paper called '*Method of Symmetrical Coordinates Applied to the Solution of Polyphase Networks*' in the Transactions of the American Institute of Electrical Engineers. In this paper, he proposed a method to resolve an unbalanced set of $n$-phasors into $n - 1$ balanced $n$-phase systems of different phase sequence and one zero-phase system in which all phasors are of equal magnitude and angle. This approach will be illustrated for a three-phase system. Figure 2.1 shows three such sets of symmetrical components.

$$\begin{aligned} V_a &= V_{a1} + V_{a2} + V_{a0} \\ V_b &= V_{b1} + V_{b2} + V_{b0} \\ V_c &= V_{c1} + V_{c2} + V_{c0} \end{aligned} \tag{2.1}$$

where $V_a$, $V_b$, $V_c$ are three phasors that are not in balance and $V_{a1}$, $V_{b1}$, $V_{c1}$ and $V_{a2}$, $V_{b2}$, $V_{c2}$ are two sets of three balanced phasors with an angle of 120° between the components $a$, $b$, and $c$. The components of the phasor set $V_{a0}$, $V_{b0}$, $V_{c0}$ are identical in amplitude and angle. Equation (2.1) can be simplified by making use of the $a$-operator:

$$a = e^{j2\pi/3}$$

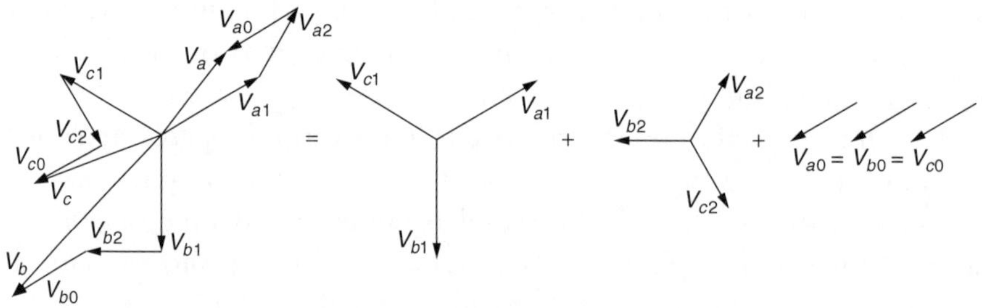

**Figure 2.1** A set of three unbalanced voltage phasors resolved in three sets of symmetrical components

The relation between the set of phasors ($V_a$, $V_b$, $V_c$) and the positive phasors, negative phasors, and zero phasors is

$$\begin{bmatrix} V_a \\ V_b \\ V_c \end{bmatrix} = \begin{bmatrix} 1 & 1 & 1 \\ 1 & a^2 & a \\ 1 & a & a^2 \end{bmatrix} \begin{bmatrix} V_{a0} \\ V_{a1} \\ V_{a2} \end{bmatrix} \tag{2.2}$$

or

$$\boldsymbol{V}_{abc} = \boldsymbol{A}\boldsymbol{V}_{012} \tag{2.3}$$

The $a$-operator rotates any phasor quantity by $120^\circ$ and the inverse relation of Equation (2.2) can be written as

$$\begin{bmatrix} V_{a0} \\ V_{a1} \\ V_{a2} \end{bmatrix} = \frac{1}{3} \begin{bmatrix} 1 & 1 & 1 \\ 1 & a & a^2 \\ 1 & a^2 & a \end{bmatrix} \begin{bmatrix} V_a \\ V_b \\ V_c \end{bmatrix} \tag{2.4}$$

or

$$\boldsymbol{V}_{012} = \boldsymbol{A}^{-1}\boldsymbol{V}_{abc} \tag{2.5}$$

In Equation (2.4), 0 refers to the *zero sequence*, 1 to the *positive sequence*, and 2 to the *negative sequence*. The names *zero, positive*, and *negative* refer to the sequence of rotation of the phasors. The positive-sequence set of phasors ($V_{a1}$, $V_{b1}$, $V_{c1}$) is the same as the voltages produced by a synchronous generator in the power system that has phase sequence a-b-c. The negative sequence ($V_{a2}$, $V_{b2}$, $V_{c2}$) has phase sequence a-c-b. The zero sequence phasors ($V_{a0}$, $V_{b0}$, $V_{c0}$) have zero-phase displacement and are identical. The symmetrical component transformation is unique if the matrix operator $\boldsymbol{A}$ is nonsingular. If $\boldsymbol{A}$ is nonsingular, its inverse $\boldsymbol{A}^{-1} = \boldsymbol{A}$ exists. The method of symmetrical components applies to any set of unbalanced three-phase quantities; similarly, for currents we have relations identical to Equation (2.4) and Equation (2.5).

## 2.2 SEQUENCE COMPONENTS FOR UNBALANCED NETWORK IMPEDANCES

A *general* three-phase system has unequal self-impedances and mutual impedances, as depicted in Figure 2.2:

$$\begin{aligned} Z_{aa} \neq Z_{bb} \neq Z_{cc} \\ Z_{ab} \neq Z_{bc} \neq Z_{ca} \end{aligned} \tag{2.6}$$

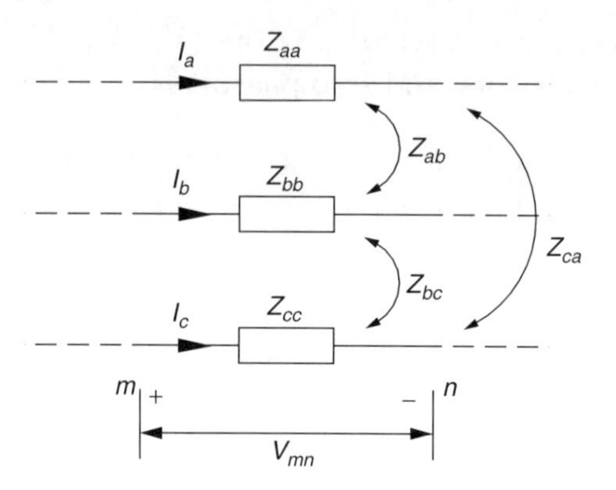


**Figure 2.2** A general three-phase system

Both the self-impedances and mutual impedances constitute sets of unbalanced or unequal complex impedances, and even balanced currents produce unequal voltage drops between $m$ and $n$. The voltage-drop equation from $m$ to $n$ can be written in matrix form as

$$V_{mn} = \begin{bmatrix} V_{mn-a} \\ V_{mn-b} \\ V_{mn-c} \end{bmatrix} = \begin{bmatrix} Z_{aa} & Z_{ab} & Z_{ac} \\ Z_{ba} & Z_{bb} & Z_{bc} \\ Z_{ca} & Z_{cb} & Z_{cc} \end{bmatrix} \begin{bmatrix} I_a \\ I_b \\ I_c \end{bmatrix} \tag{2.7}$$

By applying the symmetrical components transform to both sides, we get

$$\boldsymbol{AV}_{mn-012} = \boldsymbol{ZAI}_{012} \tag{2.8}$$

The symmetrical components of the voltage drop are given by

$$\mathbf{V}_{mn-012} = \boldsymbol{A}^{-1}\boldsymbol{ZAI}_{012} = \boldsymbol{Z}_{mn-012}\boldsymbol{I}_{012} \tag{2.9}$$

$Z$ is a transform that takes a current vector $\boldsymbol{I}_{abc}$ into a voltage-drop vector $\boldsymbol{V}_{mn}$, both in the a-b-c system. $\boldsymbol{A}$ is a linear operator that transforms currents and voltages from the 0-1-2 coordinate system into the a-b-c system. The new impedance matrix $\boldsymbol{Z}_{mn-012}$ can be found directly (see Equation (2.9)):

$$Z_{mn-012} = \begin{bmatrix} (Z_{s0} + 2Z_{m0}) & (Z_{s2} - Z_{m2}) & (Z_{s1} - Z_{m1}) \\ (Z_{s1} - Z_{m1}) & (Z_{s0} - Z_{m0}) & (Z_{s2} + 2Z_{m2}) \\ (Z_{s2} - Z_{m2}) & (Z_{s1} + 2Z_{m1}) & (Z_{s0} - Z_{m0}) \end{bmatrix} \tag{2.10}$$

with

$$\begin{aligned} Z_{s0} &= \tfrac{1}{3}(Z_{aa} + Z_{bb} + Z_{cc}) \\ Z_{s1} &= \tfrac{1}{3}(Z_{aa} + aZ_{bb} + a^2 Z_{cc}) \\ Z_{s2} &= \tfrac{1}{3}(Z_{aa} + a^2 Z_{bb} + aZ_{cc}) \end{aligned} \tag{2.11}$$

and

$$\begin{aligned} Z_{m0} &= \tfrac{1}{3}(Z_{bc} + Z_{ca} + Z_{ab}) \\ Z_{m1} &= \tfrac{1}{3}(Z_{bc} + aZ_{ca} + a^2 Z_{ab}) \\ Z_{m2} &= \tfrac{1}{3}(Z_{bc} + a^2 Z_{ca} + aZ_{ab}) \end{aligned} \tag{2.12}$$

We made use of the property of the $a$-operator $1 + a + a^2 = 0$ and $a^3 = 1$. In Equation (2.12), we made use of the property that mutual impedances of passive networks are reciprocal, and in this case it means that $Z_{ab} = Z_{ba}$, $Z_{ac} = Z_{ca}$ and so forth. When the impedance matrix of Equation (2.10) is substituted in Equation (2.9), the equation for the positive-sequence component of the voltage drop $\mathbf{V}_{mn-012}$ is

$$V_{mn-1} = (Z_{s1} - Z_{m1})I_{a0} + (Z_{s0} - Z_{m0})I_{a1} + (Z_{s2} + 2Z_{m2})I_{a2} \tag{2.13}$$

The positive-sequence voltage drop depends not only on $I_{a1}$ but also on $I_{a2}$, and this means that there is a mutual coupling between the sequences. Further, we can conclude that $\mathbf{Z}_{mn-012}$ is not symmetric; therefore, the mutual effects are not reciprocal and this is a rather disturbing result. This is the reason that we prefer to work with the special cases of both self-impedances and mutual impedances in which the matrix $\mathbf{Z}_{mn-012}$ is simplified. In many practical cases, the mutual impedances can be neglected because they are small compared with the self-impedances. The matrix $\mathbf{Z}_{mn-012}$, however, is nonsymmetric with respect to $Z_s$ and $Z_m$ terms and is therefore not made symmetric by eliminating either the self-terms or the mutual terms, and because elimination of self-impedance terms cannot be applied (because of the inherent nature of the power system), a simplification must be sought in the special case of *equal impedance* and *symmetric impedance*. In many practical power system problems, the self-impedances or mutual impedances are equal in all the three phases. In such cases, Equation (2.11) and Equation (2.12) become

$$Z_{s0} = Z_{aa}, \quad Z_{s1} = Z_{s2} = 0 \tag{2.14}$$

and

$$Z_{m0} = Z_{bc}, \quad Z_{m1} = Z_{m2} = 0 \tag{2.15}$$

If we substitute Equation (2.14) and Equation (2.15) in Equation (2.10) and examine the result, we see that the off-diagonal terms of $\mathbf{Z}_{mn-012}$ are eliminated, that the impedance matrix $\mathbf{Z}_{mn-012}$ has become reciprocal, and that zero-coupling exists between the sequences.

A less-restrictive case than that of equal impedances is the one in which the self-impedances or mutual impedances are symmetric with respect to one phase, for example, for phase a:

$$Z_{bb} = Z_{cc} \text{ and } Z_{ab} = Z_{ca} \tag{2.16}$$

In this case, the self-impedances become

$$\begin{aligned} Z_{s0} &= \tfrac{1}{3}(Z_{aa} + 2Z_{bb}) \\ Z_{s1} &= Z_{s2} = \tfrac{1}{3}(Z_{aa} - Z_{bb}) \end{aligned} \tag{2.17}$$

and the mutual impedances are

$$\begin{aligned} Z_{m0} &= (Z_{bc} + 2Z_{ab}) \\ Z_{m1} &= Z_{m2} = (Z_{bc} - Z_{ab}) \end{aligned} \tag{2.18}$$

When $\mathbf{Z}_{mn-012}$ is *diagonal*, it means that the sequences are uncoupled and currents from one sequence produce voltage drops only in that sequence – this is a very desirable characteristic. A symmetric impedance matrix means that there is a mutual coupling between sequences but that it is the reciprocal; the coupling from positive to negative sequences is exactly the same as the coupling from negative to positive. This situation can be simulated by a passive network. A *nonsymmetric* impedance matrix means that the mutual coupling is not the same between two sequences; this situation requires, in general, controlled voltage sources but its mathematical representation is no more difficult than that in the *symmetric* case. It requires computation of all matrix elements instead of computing only the upper or lower triangular matrix, as in the symmetric case. In most of the practical power system calculations, the self-impedances are considered to be equal, and except for the case of nonsymmetric mutual impedances, the problem is one of a diagonal or a symmetric matrix representation.

## 2.3 THE SEQUENCE NETWORKS

In the case of an unbalanced load or a fault supplied from balanced or equal-phase impedances, the unbalanced portion of the physical power system can be isolated for study, the rest of the power system being considered as balanced. In such a case, we determine the symmetrical components of voltage and current at the point of unbalance and transform them to determine the system a-b-c quantities. Therefore, the major objective in problem-solving is to find the sequence quantities, and for this purpose, sequence networks are introduced.

The *fault point* of a power system is that point to which the unbalanced connection is attached in the otherwise balanced system. For example, a single line-to-ground fault at bus $M$ makes bus $M$ the fault point of the system and an unbalanced three-phase load at bus $N$ defines $N$ as the fault point. In general terminology, a fault must be interpreted as any connection or situation that causes an unbalance among the three phases of the power system.

A *sequence network* is a copy of the original balanced power system to which the fault point is connected and which contains the same per-phase impedances as the physical, balanced power system. The value of each impedance is a value unique to each sequence and it can be determined by applying Thevenin's theorem by considering the sequence network to be a two-terminal or a one-port network. Because the positive- and negative-sequence currents are both balanced three-phase current sets, they see the same impedance in a passive three-phase network. The zero currents, however, generally see an impedance that is different from the positive- and negative-sequence impedance. Care must be taken when the machine impedance, of the supplying synchronous generators or of the asynchronous motors in certain loads, has to be taken into account, because the sequence impedances for electrical machines are usually different for all the three sequences!

Sequence networks are drawn as boxes in which the fault point $F$, the zero-potential bus $N$ (often the neutral connection), and the Thevenin voltage are shown. Figure 2.3 shows the sequence networks for the zero, positive, and negative sequences.

By definition, the direction of the sequence current is away from the $F$ terminal. This is because the unbalanced connection is to be attached

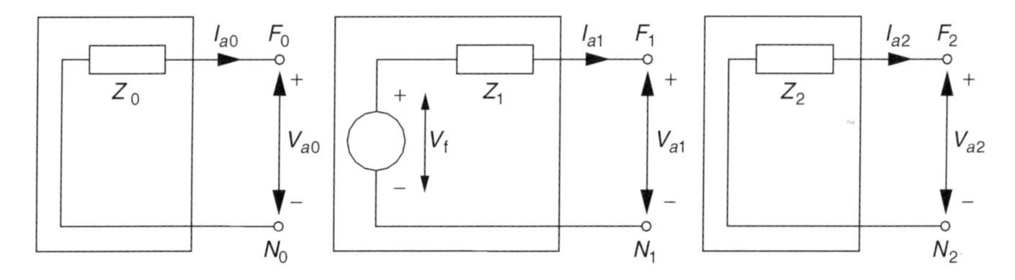


Figure 2.3 Sequence networks for the zero, positive, and negative sequences

at $F$, external to the sequence networks, and the currents are assumed to flow toward this unbalanced connection. The polarity of the voltage is defined to be a rise from $N$ to $F$ – this makes $V_{a1}$ positive for a normal power system. The Thevenin equivalent voltage $V_f$ in the positive-sequence network is the voltage of phase a at the fault point $F$ before the fault occurred. The relation for the voltage drop from $F$ to $N$ is an important one and from Figure 2.3, we can write this voltage drop in a matrix notation as

$$\begin{bmatrix} V_{a0} \\ V_{a1} \\ V_{a2} \end{bmatrix} = \begin{bmatrix} 0 \\ V_f \\ 0 \end{bmatrix} - \begin{bmatrix} Z_0 & 0 & 0 \\ 0 & Z_1 & 0 \\ 0 & 0 & Z_2 \end{bmatrix} \begin{bmatrix} I_{a0} \\ I_{a1} \\ I_{a2} \end{bmatrix} \tag{2.19}$$

## 2.4 *THE ANALYSIS OF UNSYMMETRICAL THREE-PHASE FAULTS*

The symmetrical components and, in particular, the sequence networks are very useful tools when dealing with unsymmetrical faults. The analysis of unsymmetrical faults is rather straightforward. First the three-phase circuit of the fault is drawn and currents, voltages, and impedances are labelled, taking into consideration the directions and polarities. Then the *boundary conditions* for the unsymmetrical fault conditions are determined in the a-b-c system, and these current and voltage relations are transformed from the a-b-c system to the 0-1-2 coordinate system by using the $A$ or $A^{-1}$ transformation. Next, the $F$- and $N$-terminals of the sequence networks are connected so that the sequence currents comply with the boundary conditions in the 0-1-2 system. The boundary conditions for the sequence voltages determine the connection of the remaining terminals of the sequence networks. These rather straightforward steps are illustrated by the analysis of two power system fault cases that are taken as typical for circuit breaker testing and standardisation. The single line-to-ground fault is a very common fault type; much less common is a three-phase-to-ground fault. The three-phase-to-ground fault, however, is a severe fault to be interrupted by high-voltage circuit breakers.

### 2.4.1 The Single Line-to-Ground Fault

The three-phase circuit of the single line-to-ground fault (SLG) is drawn in Figure 2.4.

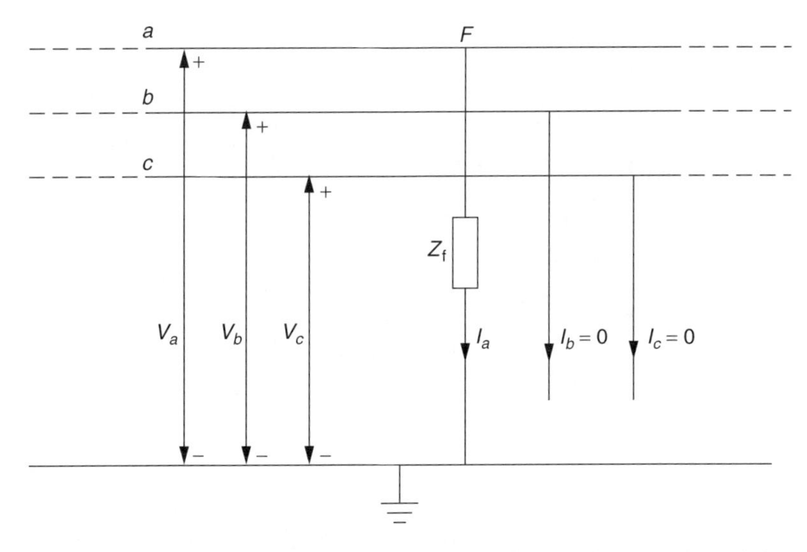


**Figure 2.4** Three-phase circuit diagram of an SLG fault at fault point F of the power system

The boundary conditions in the a-b-c system can be derived by inspection of Figure 2.4:

$$I_b = I_c = 0 \tag{2.20}$$

$$V_a = Z_f I_a \tag{2.21}$$

These boundary equations are *transformed* from the a-b-c system to the 0-1-2 coordinate system:

$$I_{012} = A^{-1} I_{abc} \tag{2.22}$$

$$I_{012} = \frac{1}{3}\begin{bmatrix} 1 & 1 & 1 \\ 1 & a & a^2 \\ 1 & a^2 & a \end{bmatrix}\begin{bmatrix} I_a \\ 0 \\ 0 \end{bmatrix} = \frac{1}{3} I_a \begin{bmatrix} 1 \\ 1 \\ 1 \end{bmatrix} \tag{2.23}$$

This implies that all the sequence currents are equal, and when the boundary equations for the voltage relations of Equation (2.21) are transformed, we find the relation

$$V_{a0} + V_{a1} + V_{a2} = Z_f I_a = 3 Z_f I_{a1} \tag{2.24}$$

The fact that the sequence currents are equal implies that the sequence networks must be connected in series. From Equation (2.24), we note that the sequence voltages add to $3Z_f I_{a1}$ – this requires the addition of

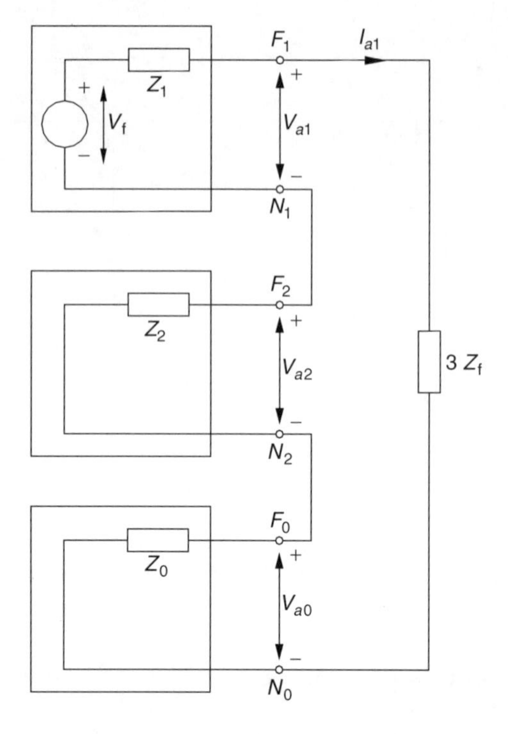


**Figure 2.5** Sequence network connection for an SLG fault

an external impedance. The connection of the sequence networks for a single line-to-ground fault is depicted in Figure 2.5. From Figure 2.5, we can write

$$I_{a0} = I_{a1} = I_{a2} = \frac{V_f}{Z_0 + Z_1 + Z_2 + 3Z_f} \tag{2.25}$$

and now that the sequence current relations are known, we can determine the sequence voltage relations from Equation (2.24).

### 2.4.2 The Three-Phase-To-Ground Fault

The three-phase-to-ground fault is, in fact, a symmetrical fault because the power system remains in balance after the fault occurs. It is the most severe fault type and other faults, if not cleared promptly, can easily develop into it. The *three-phase circuit* of the three-phase-to-ground fault (TPG) is drawn in Figure 2.6.

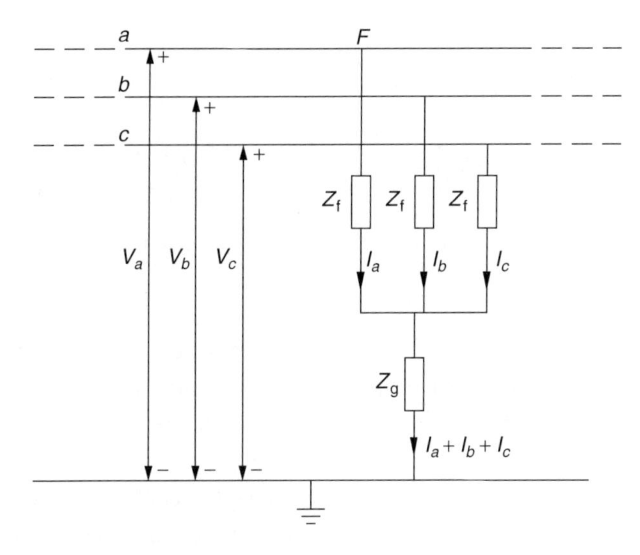


**Figure 2.6** Three-phase circuit diagram of a TPG fault at fault point $F$ of the power system

The boundary conditions in the a-b-c system can be derived by inspection of Figure 2.6:

$$V_a = Z_f I_a + Z_g(I_a + I_b + I_c) \tag{2.26}$$

$$V_b = Z_f I_b + Z_g(I_a + I_b + I_c) \tag{2.27}$$

$$V_c = Z_f I_c + Z_g(I_a + I_b + I_c) \tag{2.28}$$

The boundary conditions are again transformed from the a-b-c system to the 0-1-2 coordinate system, and when we write in terms of the symmetrical components of phase a, we get

$$V_a = (V_{a0} + V_{a1} + V_{a2}) = Z_f(I_{a0} + I_{a1} + I_{a2}) + 3Z_g I_{a0} \tag{2.29}$$

$$V_b = (V_{a0} + a^2 V_{a1} + aV_{a2}) = Z_f(I_{a0} + a^2 I_{a1} + aI_{a2}) + 3Z_g I_{a0} \tag{2.30}$$

$$V_c = (V_{a0} + aV_{a1} + a^2 V_{a2}) = Z_f(I_{a0} + aI_{a1} + a^2 I_{a2}) + 3Z_g I_{a0} \tag{2.31}$$

It is considered that $I_a + I_b + I_c = 3I_{a0} = 0$ because the fault impedances $Z_f$ and also the supply voltages in each phase are in balance. Therefore, the currents are also in balance and we can write

$$I_{012} = \frac{1}{3}\begin{bmatrix} 1 & 1 & 1 \\ 1 & a & a^2 \\ 1 & a^2 & a \end{bmatrix}\begin{bmatrix} I_a \\ I_b \\ I_c \end{bmatrix} = \begin{bmatrix} 0 \\ I_{a1} \\ 0 \end{bmatrix} \tag{2.32}$$

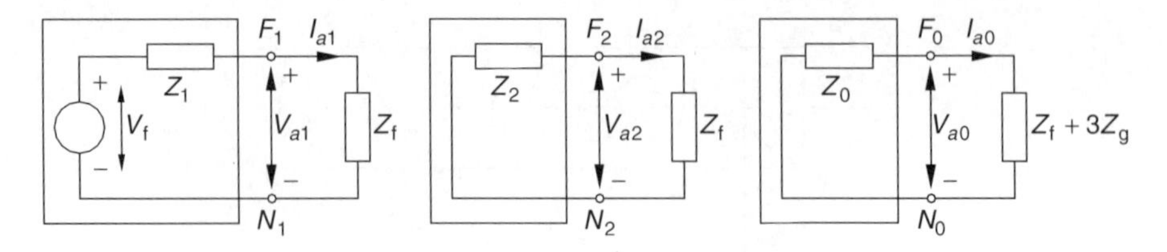


**Figure 2.7** Connection of the sequence network for a TPG fault

This leads to $I_{a1} = I_a$, $I_{a0} = I_{a2} = 0$. The sequence networks are therefore connected as shown in Figure 2.7. When the fault impedance $Z_f$ is small, or even zero, the TPG fault is in fact a short circuit.

After clearing a three-phase fault, the power system changes from the steady-state situation, in which the three-phase short-circuit current is flowing, to the state in which only the power frequency–recovery voltage is present across the contacts of the circuit breaker. In an inductive circuit, the change from one steady state to another is always accompanied by a transient – the transient recovery voltage or TRV. At current zero, the arc voltage and the arc current extinguish and the TRV oscillates from zero to the crest of the AC power frequency–recovery voltage (See Chapter 4, *Circuit Breakers*). One of the breaker poles clears first and the two last clearing poles interrupt 90° later – they, in fact, form a single-phase circuit. Of interest is the value of the AC power frequency–recovery voltage of the first clearing phase because the TRV oscillates to this value. Let us consider the situation depicted in Figure 2.8.

The system is grounded by means of a neutral impedance at the star point of the three-phase delta/wye transformer. When the first pole of the circuit breaker interrupts the short-circuit current, let us assume that this is phase a, the other two poles are still arcing and therefore in a conducting state. In fact, these two last clearing poles interrupt a *double line-to-ground*

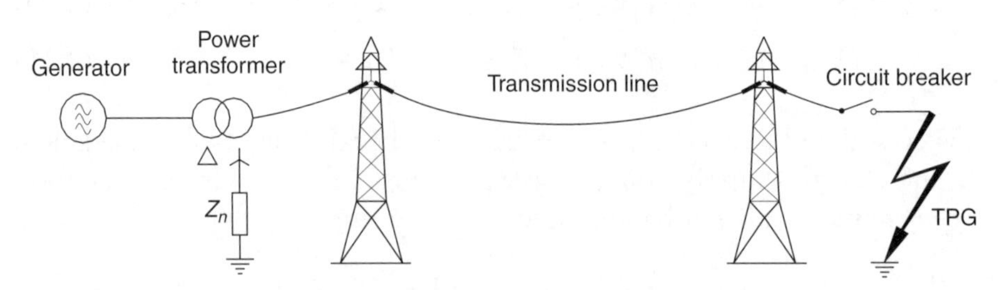


**Figure 2.8** The interruption of a three-phase line-to-ground-fault in a power system grounded via a neutral impedance

(DLG) fault. The boundary conditions for this DLG fault are

$$\begin{aligned} I_a &= 0 \\ V_b &= (Z_f + Z_g)I_b + Z_g I_c \\ V_c &= (Z_f + Z_g)I_c + Z_g I_b \end{aligned} \tag{2.33}$$

These boundary conditions are *transformed* from the a-b-c system to the 0-1-2 coordinate system and this results in

$$I_a = 0 = I_{a0} + I_{a1} + I_{a2} \tag{2.34}$$

From Equation (2.2) we can write

$$V_b = V_{a0} + a^2 V_{a1} + a V_{a2} \tag{2.35}$$

$$V_c = V_{a0} + a V_{a1} + a^2 V_{a2} \tag{2.36}$$

and for the difference,

$$V_b - V_c = -j\sqrt{3}(V_{a1} - V_{a2}) \tag{2.37}$$

From Equation (2.33) we can also write

$$V_b - V_c = Z_f(I_b - I_c) \tag{2.38}$$

Substituting Equation (2.37) into Equation (2.38) gives

$$-j\sqrt{3}(V_{a1} - V_{a2}) = Z_f(I_b - I_c) \tag{2.39}$$

$$(V_{a1} - V_{a2}) = Z_f\left(\frac{I_b - I_c}{-j\sqrt{3}}\right) = Z_f(I_{a1} - I_{a2}) \tag{2.40}$$

or

$$V_{a1} - Z_f I_{a1} = V_{a2} - Z_f I_{a2} \tag{2.41}$$

Adding Equation (2.35) and Equation (2.36) results in

$$V_b + V_c = 2V_{a0} - (V_{a1} + V_{a2}) \tag{2.42}$$

and adding $V_b$ and $V_c$ from Equation (2.33) gives us

$$V_b + V_c = Z_f(I_b + I_c) + 2Z_g(I_b + I_c) \tag{2.43}$$

$$I_b = I_{a0} + a^2 I_{a1} + a I_{a2}$$
$$I_c = I_{a0} + a I_{a1} + a^2 I_{a2} \tag{2.44}$$
$$I_b + I_c = 2I_{a0} + (a + a^2)I_{a1} + (a + a^2)I_{a2} = 2I_{a0} - I_{a1} - I_{a2} \tag{2.45}$$

Substituting Equation (2.45) in Equation (2.43), we get

$$V_b + V_c = Z_f[2I_{a0} - (I_{a1} + I_{a2})] + Z_g[4I_{a0} - 2(I_{a1} + I_{a2})] \tag{2.46}$$

Because Equation (2.42) and Equation (2.46) are equal, we can collect terms and write

$$2V_{a0} - 2Z_f I_{a0} - 4Z_g I_{a0} = V_{a1} + V_{a2} - Z_f(I_{a1} + I_{a2}) - 2Z_g(I_{a1} + I_{a2}) \tag{2.47}$$

By using Equation (2.41) and the property that $I_{a1} + I_{a2} = -I_{a0}$, we find, after rearranging, that

$$V_{a0} - Z_f I_{a0} - 3Z_g I_{a0} = V_{a1} - Z_f I_{a1} \tag{2.48}$$

From Equation (2.34), we see immediately that the $N$-terminals of the sequence networks must be connected to a common node. Equation (2.41) shows us that the voltages across the positive and negative sequence networks are equal if an external impedance $Z_f$ is added in series with each network. These conditions are met when all the three sequence networks are connected in parallel, as shown in Figure 2.9.

From the inspection of the parallel connection of the three sequence networks, it follows that

$$I_{a1} = \frac{V_f}{Z_1 + Z_f + \dfrac{(Z_2 + Z_f)(Z_0 + Z_f + 3Z_g)}{Z_0 + Z_2 + 2Z_f + 3Z_g}} \tag{2.49}$$

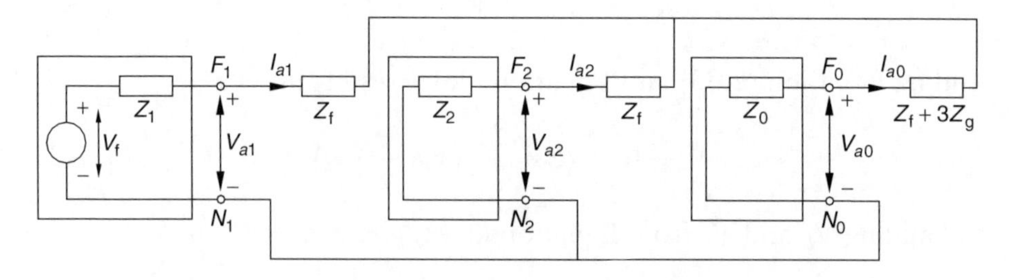


**Figure 2.9** Connection of the sequence networks for a DLG fault

and

$$V_{a1} = V_f - Z_1 I_{a1} \tag{2.50}$$

In the case of a bolted fault to ground, $Z_f = 0$ and $Z_g = 0$, and the transformation back to the a-b-c domain by means of the Equation (2.3) and the use of the boundary condition that $V_{a0} = V_{a1} = V_{a2}$ result in the recovery voltage across the contacts of pole a:

$$V_a = 3V_{a1} = 3V_f \frac{Z_2 Z_0/(Z_2 + Z_0)}{\left(Z_1 + \dfrac{Z_2 Z_0}{(Z_2 + Z_0)}\right)} \tag{2.51}$$

The ratio between the voltage across the first clearing pole and the phase voltage of the undistorted power system is called the *first-pole-to-clear* (FPTC) factor

$$\text{FPTC} = \frac{V_a}{V_f} = 3\frac{Z_2 Z_0}{Z_1 (Z_2 + Z_0) + Z_2 Z_0} \tag{2.52}$$

The positive-, negative-, and zero-sequence impedances $Z_1$, $Z_2$, and $Z_0$ have a resistive and an inductive component $Z = R + jX$. Under normal system conditions, the load determines the power factor, but during a short circuit, the load is short-circuited and the system is mainly inductive. Therefore, we can put for $Z_1$, $Z_2$, and $Z_0$ the inductance values $X_1$, $X_2$, and $X_0$, respectively.

When the fault is relatively far away from the supplying generators, the positive and negative impedances are equal and we can put $X_1 = X_2 = X$. In addition, the neutral connection of the transformer star point is a complex impedance $Z_n = R_n + jX_n$ and the zero-sequence impedance becomes $Z_0 = jX_0 + 3Z_n$. When this is substituted in Equation (2.52), we can write for the FPTC factor

$$\text{FPTC} = 3\frac{3R_n + j(X_0 + 3X_n)}{jX + 2[3R_n + j(X_0 + 3X_n)]} \tag{2.53}$$

In the case of an *ungrounded neutral*, the value of the impedance $Z_n$ is infinite and the FPTC factor is 1.5. This is an important result because a considerable number of power systems have an isolated neutral. In the case of clearing a three-phase-to-ground fault, the *peak* of the TRV can, when there is little or no damping in the system, become as high as two times the crest value of the power frequency–recovery voltage; in systems with an isolated neutral, this is two times the FPTC factor, which is three times the phase voltage.

For solidly grounded systems, $Z_n = 0$, and the expression for the FPTC factor becomes

$$\text{FPTC} = \frac{3X_0}{X + 2X_0} \tag{2.54}$$

When $X$ has a larger value then $X_0$, the FPTC factor is smaller than one. For cases in which $X = X_0$, the FPTC factor for solidly grounded systems is one, and in an undamped situation, the peak of the TRV for the first clearing pole is twice the crest value of the system phase voltage.

## *2.5 REFERENCES FOR FURTHER READING*

Anderson, P. M., *Analysis of Faulted Power Systems*, Power Systems Engineering Series, IEEE Press, 1995.

Fortescue, C. L., "Method of symmetrical coordinates applied to the solution of polyphase networks," *Trans. AIEE*, **37**(Part II), 1329 ff. (1918).

Greenwood, A., *Electrical Transients in Power Systems*, Chapter 6, Wiley & Sons, New York, 1991.

Grainger, J. J. and Stevenson, Jr., W. D., *Power System Analysis*, Chapters 11 and 12, McGraw-Hill, New York, 1994.

Peterson, H. A., *Transients in Power Systems*, Chapter 4, Section 6, Wiley & Sons, New York, Chapman & Hall, London, 1951.

Slamecka, E. and Waterschek, W., *Schaltvorgaenge in Hoch- und Niederspannungsnetzen*, Chapters 1 and 2, Siemens Aktiengesellschaft, Berlin, 1972.

Taylor, E. O., ed., *Power System Transients*, Chapter 1, Section: Faults, George Newnes, 1954.

# 3

# Travelling Waves

Power systems are large and complex systems, but for steady-state analysis, the wavelength of the sinusoidal currents and voltages is still large compared with the physical dimensions of the network – for 50-Hz power frequency, the wavelength is 6000 km. For steady-state analysis, a lumped-element representation is adequate for most cases. However, for transient analysis, this is no longer the case and the travel time of the electromagnetic waves has to be taken into account. A lumped representation of, for instance, an overhead transmission line by means of pi-sections does not account for the travel time of the electromagnetic waves, as can be easily seen from Figure 3.1.

When we make a representation of an overhead transmission line or a high-voltage cable by means of a number of pi-sections, we take the properties of the electric field in a capacitance and the properties of the magnetic field in an inductance into account and connect these elements with lossless wires.

When the switch S is closed, a current flows through the first inductance $L_1$ and charges the first capacitance $C_1$. The accumulation of charge on $C_1$ creates a voltage that causes a current to flow through $L_2$. This current charges $C_2$, a voltage builds up across $C_2$, a current flows through $L_3$ and so on. This kind of reasoning shows that a disturbance at one end of the pi-section network is immediately noticeable at the other end of the network. We know from experience that this is not what happens when a source is switched on in a transmission line; it takes a certain time before the current and voltage waves reach the end of the line.

A representation of overhead lines and underground cables by means of lumped elements is not helpful in making us understand the wave phenomena because electromagnetic waves have a travel time. Only when the physical dimensions of a certain part of the power system are small compared with the wavelength of the transients, the travel time of the

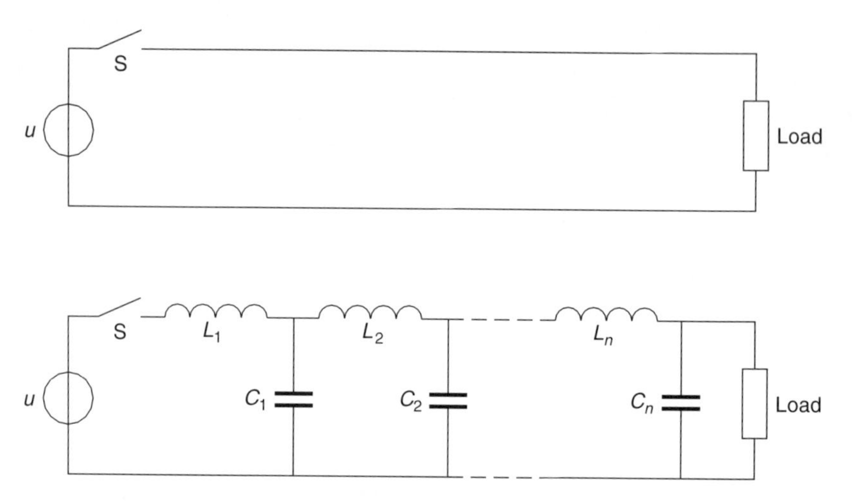


Figure 3.1 Lumped-element representation of a two-wire transmission line

electromagnetic waves can be neglected and a lumped-element representation of that particular part of the power system can be used for in-depth analysis.

If the travel time of the current and voltage waves is taken into account and we want to represent the properties of the electric field by means of a capacitance and the properties of the magnetic field by an inductance, we call the capacitance and the inductance *distributed*. An overhead transmission line, a bus bar, or an underground cable have certain physical dimensions and therefore their overall inductance, capacitance, and resistance is regarded to be equally distributed over their size. When we work with distributed elements, we must realize that the currents and voltages in, for instance, a line do not necessarily have the same value over the entire length of the line at the same instant of time.

## 3.1 VELOCITY OF TRAVELLING WAVES AND CHARACTERISTIC IMPEDANCE

If a voltage source $u$ is switched on in a two-wire transmission line at $t = 0$ (see Figure 3.2), the line will be charged by the voltage source. After a small time span $\Delta t$, only a small segment $\Delta x$ of the line will be charged instantaneously with a charge $\Delta Q = C\Delta xu$. This charge, equally distributed over the line segment $\Delta x$, causes an electric field $E$ around this line segment and the current, or the flow of charge creates a magnetic field $H$ around the line segment $\Delta x$.

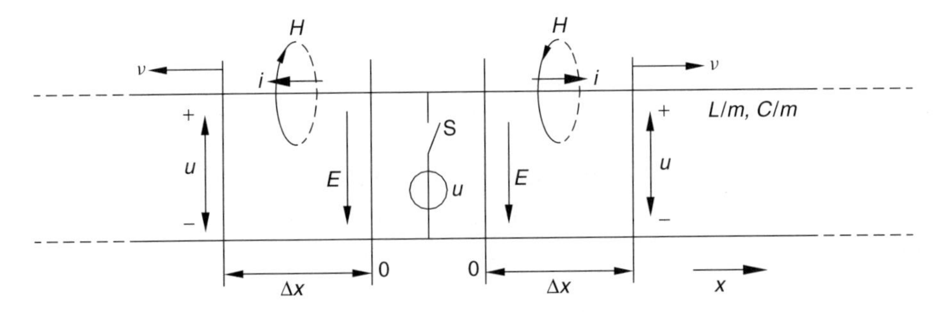


**Figure 3.2** Electric and magnetic field around a line segment $\Delta x$ of a two-wire transmission line

If $\Delta x$ is made infinitely small, the expression for the current is

$$i = \lim_{\Delta x \to 0} \frac{\Delta Q}{\Delta t} = \lim_{\Delta x \to 0} \frac{Cu\Delta x}{\Delta t} = Cu\frac{\mathrm{d}x}{\mathrm{d}t} = Cuv \qquad (3.1)$$

Because the distance $\Delta x$ is covered in the time $\Delta t$, $\Delta x/\Delta t$ is the velocity at which the charge travels along the line. The magnetic flux present around the line segment is $\Delta\Phi = L\Delta xi$. If this is substituted in Equation (3.1) and $\Delta x$ is made infinitely small, the expression for the induced electromotive force *emf* in the loop enclosed by the two wires over the distance $\Delta x$ is

$$\text{emf} = \lim_{\Delta x \to 0} \frac{\Delta \Phi}{\Delta t} = LCu\left(\frac{\mathrm{d}x}{\mathrm{d}t}\right)^2 = LCuv^2 \qquad (3.2)$$

Because there cannot be a discontinuity in voltage, this emf equals the voltage source $u$; this gives an expression for the wave velocity:

$$v = \frac{1}{\sqrt{LC}} \qquad (3.3)$$

The wave velocity depends only on the geometry of the line and on the permittivity and the permeability of the surrounding medium.

For a 150-kV overhead transmission line with one conductor per phase, a conductor radius of $r = 25$ mm and an equivalent distance between phases of $d_m = 5.5$ m, the inductance and the capacitance values are

$$L = \left(\frac{\mu_0}{2\pi}\right)\ln\left(\frac{d_m}{0.779r}\right) = 1.13 \text{ mH/km} \qquad (3.4)$$

$$C = \frac{2\pi\varepsilon_0}{\ln\left(\frac{d_m}{r}\right)} = 10.3 \text{ nF/km} \tag{3.5}$$

This results in a wave velocity of $\nu_{\text{line}} = 293.117$ km/s. When we calculate the distributed inductance and the capacitance for a single-core cross-linked polyethylene (XLPE)–insulated 18/30-kV medium-voltage cable with a 120-mm$^2$ copper conductor and a 16-mm$^2$ earth screen (N2XS2Y-1x120/16-18/30 kV), the values are

$$L = 0.430 \text{ mH/km} \qquad C = 0.178\ \mu\text{F/km}$$

This results in a wave velocity of $\nu_{\text{cable}} = 114.302$ km/s. On an overhead transmission line the electromagnetic waves propagate close to the speed of light, but in an underground cable the velocity is considerably lower. When the wave velocity is substituted in Equation (3.1) we get

$$i = Cuv = \frac{Cu}{\sqrt{LC}} \tag{3.6}$$

We notice that the ratio between the voltage and current wave has a fixed value

$$\frac{u}{i} = \sqrt{\frac{L}{C}} \tag{3.7}$$

This is called the *characteristic impedance* of a transmission line. The characteristic impedance depends on the geometry of the transmission line and its surrounding medium only. For the 150-kV overhead line, the characteristic impedance is $Z_{\text{line}} = 330\ \Omega$, and for the XLPE medium-voltage cable, the characteristic impedance is $Z_{\text{cable}} = 49\ \Omega$.

## 3.2 ENERGY CONTENTS OF TRAVELLING WAVES

An electromagnetic wave contains energy. This energy is stored in the electric field and the magnetic field. The small line segment $\Delta x$ in Figure 3.3 has a capacitance $C\Delta x$ and is charged to a voltage $u(x_0, t_0)$. For the electric energy contents of the line segment at the time $t_0$, we can write

$$w_u = \tfrac{1}{2} C\Delta x u(x_0, t_0)^2 \tag{3.8}$$

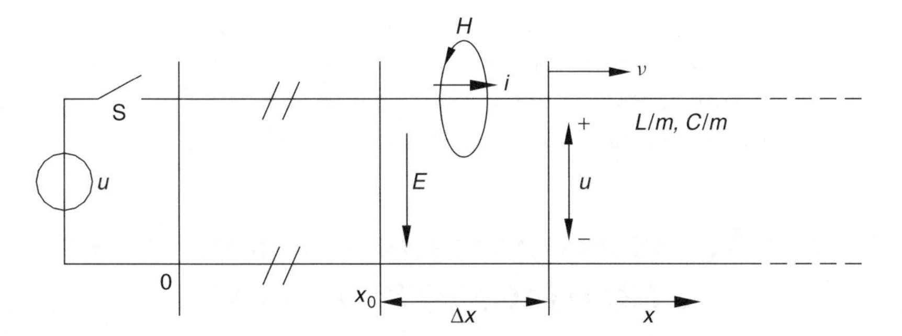


**Figure 3.3** A segment $\Delta x$ of a two-conductor line for transverse electromagnetic field mode of propagation

The magnetic energy contents of the loop enclosed by the two wires over the distance $\Delta x$ at the time $t_0$ is

$$w_i = \tfrac{1}{2} L \Delta x i(x_0, t_0)^2 \tag{3.9}$$

When the current–voltage relation $u(x, t) = (L/C)^{1/2} i(x, t)$ is substituted in the expression for the electric energy, we get the expression for the magnetic energy, and when the current–voltage relation is substituted in the expression for the magnetic energy, we get the expression for the electric energy. This means that $w_u/w_i = 1$ and that $u(x, t)^2/i(x, t)^2 = L/C = Z^2$.

The energy of an electromagnetic wave is equally distributed over the electric and the magnetic field. The total energy of line segment $\Delta x$ at the time $t_0$ is the sum of the energy in the electric and the magnetic field:

$$w_t = w_u + w_i = C \Delta x u(x_0, t_0)^2 = L \Delta x i(x_0, t_0)^2 \tag{3.10}$$

The energy passing through $\Delta x$ per second is the power $P$ of the electromagnetic wave:

$$\begin{aligned} P &= \frac{w_t}{\Delta t} = \frac{w_t}{(\Delta x / v)} = C u(x_0, t_0)^2 \frac{1}{\sqrt{LC}} = \frac{u(x_0, t_0)^2}{Z} \\ &= L i(x_0, t_0)^2 \frac{1}{\sqrt{LC}} = i(x_0, t_0)^2 Z \end{aligned} \tag{3.11}$$

The power of an electromagnetic wave can be enormous (for example, 26.7 MW for a voltage wave with a crest of 100 kV on an overhead transmission line with a characteristic impedance $Z = 375\ \Omega$), but the duration of switching and lightning transients is usually short, ranging from microseconds to milliseconds. Still, the energy contents of switching and lightning transients can easily reach values up to a few MJ and can cause considerable damage to power system equipment.

## 3.3 ATTENUATION AND DISTORTION OF ELECTROMAGNETIC WAVES

Until now we have looked at the transmission line as a so-called lossless line, which means that the following items have not been taken into account:

- The series resistance of the conductors
- The skin-effect for higher frequencies
- The losses in the dielectric medium between the conductors in a high-voltage cable
- The leakage currents across string insulators
- The influence of the ground resistance
- The corona losses and so on

To include these losses in our analysis, we assume a series resistance $R$ and a parallel conductance $G$, evenly distributed along the wires as the inductance $L$ and the capacitance $C$. We consider again a line segment $\Delta x$ (see Figure 3.4).

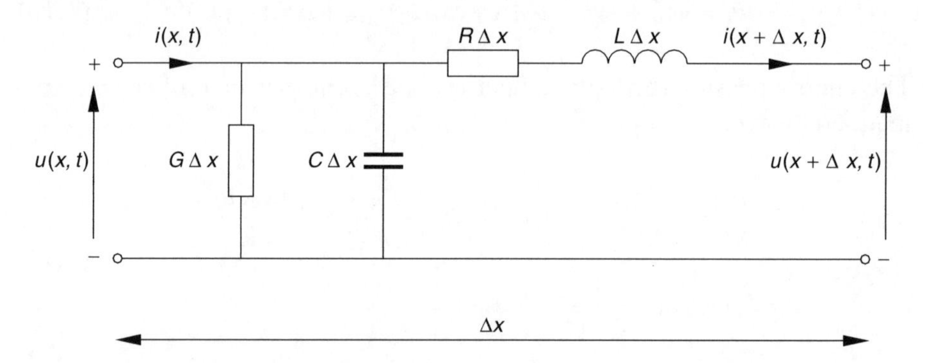


**Figure 3.4** Equivalent circuit of a differential-length line segment $\Delta x$ of a two-conductor transmission line

The losses in line segment $\Delta x$ at the time $t_0$ are

$$P_{\text{Loss}} = i(x_0, t_0)^2 R\Delta x + u(x_0, t_0)^2 G\Delta x = i(x_0, t_0)^2 (R + Z^2 G)\Delta x \quad (3.12)$$

When $\Delta x$ becomes infinitely small, we can write it as $\mathrm{d}x$. An electromagnetic wave with a power $P = i(x_0, t_0)^2 Z$ entering this infinitely small line segment $\mathrm{d}x$ loses (when we differentiate $P$ with respect to $i(x, t)$) power $\Delta P = 2i(x_0, t_0)\,\mathrm{d}iZ$, while travelling the distance $\mathrm{d}x$. This loss of power has a negative sign because the wave energy decreases and is dissipated in $R$ and $G$:

$$i(x_0, t_0)^2 (R + Z^2 G)\,\mathrm{d}x = -2i(x_0, t_0)\,\mathrm{d}iZ \quad (3.13)$$

This can be generalised for *every* instant of time

$$\frac{\mathrm{d}i}{i(x,t)} = -0.5\left(\frac{R}{Z} + ZG\right)\mathrm{d}x \Longrightarrow \frac{\mathrm{d}i}{\mathrm{d}x} = -0.5\left(\frac{R}{Z} + ZG\right)i(x,t) \quad (3.14)$$

The solution of this differential equation is

$$i(x, t) = i(x_0, t_0)\mathrm{e}^{-0.5[(R/Z)+(ZG)]x} \quad (3.15)$$

and because of the relation $u(x, t) = Zi(x, t)$ that exists between the voltage and the current wave, we can write for the voltage wave

$$u(x, t) = Zi(x_0, t_0)\mathrm{e}^{-0.5[(R/Z)+(ZG)]x} = u(x_0, t_0)\mathrm{e}^{-0.5[(R/Z)+(ZG)]x} \quad (3.16)$$

When the voltage and current waves travel along a transmission line with losses, the amplitude of the waves is exponentially decreased. This is called *attenuation* and is caused by the properties of the transmission line.

For overhead transmission lines, $G$ is a very small number and we can simplify Equation (3.15) for the current and Equation (3.16) for the voltage wave to

$$i(x, t) = i(x_0, t_0)\mathrm{e}^{-(Rx/2Z)} \quad (3.17)$$

and

$$u(x, t) = u(x_0, t_0)\mathrm{e}^{-(Rx/2Z)} \quad (3.18)$$

The attenuation is small for a line with a low resistance and/or a large characteristic impedance.

When the series resistance $R$ and the parallel conductance $G$ can be neglected, both the wave velocity and the characteristic impedance are constant and the transmission line is said to be lossless.

When $R/L = G/C$, we call the transmission line *distortionless*; the shape of the current and voltage waves is not affected and the wave velocity and the characteristic impedance are constant, similar to a lossless line. When the transmission line is not distortionless, the steepness of the wave front will decrease and the general shape of the waves will be more elongated when they travel along the line.

## 3.4 THE TELEGRAPH EQUATIONS

The transmission line equations that govern general two-conductor uniform transmission lines, including parallel-plate, two-wire lines, and coaxial lines, are called *the telegraph equations*. The general transmission line equations are named the telegraph equations because they were formulated for the first time by Oliver Heaviside (1850–1925) when he was employed by a telegraph company and used to investigate disturbances on telephone wires.

When we consider again a line segment $\Delta x$ with parameters $R$, $G$, $L$, and $C$, all per unit length, (see Figure 3.4) the line constants for segment $\Delta x$ are $R\Delta x$, $G\Delta x$, $L\Delta x$, and $C\Delta x$. The electric flux $\psi$ and the magnetic flux $\Phi$ created by the electromagnetic wave, which causes the instantaneous voltage $u(x, t)$ and current $i(x, t)$, are

$$\mathrm{d}\Psi(t) = u(x, t)C\Delta x \tag{3.19}$$

and

$$\mathrm{d}\Phi(t) = i(x, t)L\Delta x \tag{3.20}$$

Applying Kirchhoff's voltage law on the loop enclosed by the two wires over the distance $\Delta x$, we obtain

$$\begin{aligned} u(x, t) - u(x + \Delta x, t) = -\Delta u &= i(x, t)R\Delta x + \frac{\partial}{\partial t}\mathrm{d}\Phi(t) \\ &= \left(R + L\frac{\partial}{\partial t}\right)i(x, t)\Delta\, x \end{aligned} \tag{3.21}$$

In the limit, as $\Delta x \to 0$, this voltage equation becomes

$$\frac{\partial u(x,t)}{\partial x} = -L\frac{\partial i(x,t)}{\partial t} - Ri(x,t) \tag{3.22}$$

Similarly, for the current flowing through $G$ and the current charging $C$, Kirchhoff's current law can be applied:

$$\begin{aligned} i(x,t) - i(x+\Delta x,t) = -\Delta i &= u(x,t)G\Delta x + \frac{\partial}{\partial t}\,\mathrm{d}\psi(t) \\ &= \left(G + C\frac{\partial}{\partial t}\right)u(x,t)\Delta x \end{aligned} \tag{3.23}$$

In the limit, as $\Delta x \to 0$, this current equation becomes

$$\frac{\partial i(x,t)}{\partial x} = -C\frac{\partial u(x,t)}{\partial t} - Gu(x,t) \tag{3.24}$$

The negative sign in these equations is caused by the fact that when the current and voltage waves propagate in the positive $x$-direction $i(x,t)$ and $u(x,t)$ will decrease in amplitude for increasing $x$. To solve the equations, they are transformed into the Laplace domain by substituting the Heaviside operator $p = \partial/\partial t$; this leaves us with the partial differential equations

$$-\frac{\partial u(x,p)}{\partial x} = (R + pL)i(x,p) \tag{3.25}$$

$$-\frac{\partial i(x,p)}{\partial x} = (G + pC)u(x,p) \tag{3.26}$$

When we substitute $Z' = R + pL$ and $Y' = G + pC$ and differentiate once more with respect to $x$, we get the second-order partial differential equations

$$\frac{\partial^2 u(x,p)}{\partial x^2} = -Z'\frac{\partial i(x,p)}{\partial x} = Z'Y'u(x,p) = \gamma^2 u(x,p) \tag{3.27}$$

$$\frac{\partial^2 i(x,p)}{\partial x^2} = -Y'\frac{\partial u(x,p)}{\partial x} = Y'Z'i(x,p) = \gamma^2 i(x,p) \tag{3.28}$$

$$\gamma = \sqrt{(RG + (RC + GL)p + LCp^2)} = \frac{1}{\nu}\sqrt{(p+\alpha)^2 - \beta^2} \tag{3.29}$$

In this expression,

$v = (1/LC)^{1/2}$ the wave velocity; (3.30)

$\alpha = 1/2[(R/L) + (G/C)]$ the attenuation constant (of influence on the amplitude of the travelling waves); (3.31)

$\beta = 1/2[(R/L) - (G/C)]$ the phase constant (of influence on the phase shift of the travelling waves); and (3.32)

$Z = (Z'/Y')^{1/2} = (L/C)^{1/2} \times [(p + \alpha + \beta)/(p + \alpha - \beta)]^{1/2}$ the characteristic impedance. (3.33)

The solutions of Equation (3.27) and Equation (3.28) in the time domain are

$$u(x, t) = e^{\gamma x} f_1(t) + e^{-\gamma x} f_2(t) \tag{3.34}$$

and

$$i(x, t) = -\frac{1}{Z}\left[e^{\gamma x} f_1(t) - e^{-\gamma x} f_2(t)\right] \tag{3.35}$$

In these expressions, $f_1(t)$ and $f_2(t)$ are arbitrary functions, independent of $x$.

### 3.4.1 The Lossless Line

For the lossless line, the series resistance $R$ and the parallel conductance $G$ are zero, and the propagation constant and the characteristic impedance are

$$\gamma = p\sqrt{LC} = p/v \tag{3.36}$$

and

$$Z = Z_0 = \sqrt{\frac{L}{C}} \tag{3.37}$$

For the characteristic impedance of a lossless transmission line, the symbol $Z_0$ is often used. The solutions for the voltage and current waves reduce to

$$u(x, t) = e^{px/v} f_1(t) + e^{-px/v} f_2(t) \tag{3.38}$$

$$i(x, t) = -\frac{1}{Z_0}\left[e^{px/v} f_1(t) - e^{-px/v} f_2(t)\right] \tag{3.39}$$

When we apply Taylor's series to approximate a function by a series, assuming that the function can be differentiated a sufficiently number of times,

$$f(t+h) = f(t) + hf'(t) + \left(\frac{h^2}{2!}\right) f''(t) + \cdots\cdots \tag{3.40}$$

and substituting the Heaviside operator $p = \mathrm{d}/\mathrm{d}t$ results in

$$f(t+h) = \left(1 + hp + \frac{h^2}{2!}p^2 + \cdots\cdots\right) f(t) = \mathrm{e}^{hp} f(t) \tag{3.41}$$

Applying this to Equation (3.38) and Equation (3.39), we find solutions for the voltage and current waves in the time domain:

$$u(x,t) = f_1\left(t + \frac{x}{v}\right) + f_2\left(t - \frac{x}{v}\right) \tag{3.42}$$

$$i(x,t) = -\frac{1}{Z_0}\left[f_1\left(t + \frac{x}{v}\right) - f_2\left(t - \frac{x}{v}\right)\right] \tag{3.43}$$

In this expression, $f_1[t + (x/v)]$ is a function describing a wave propagating in the negative $x$-direction, usually called *the backward wave*, and

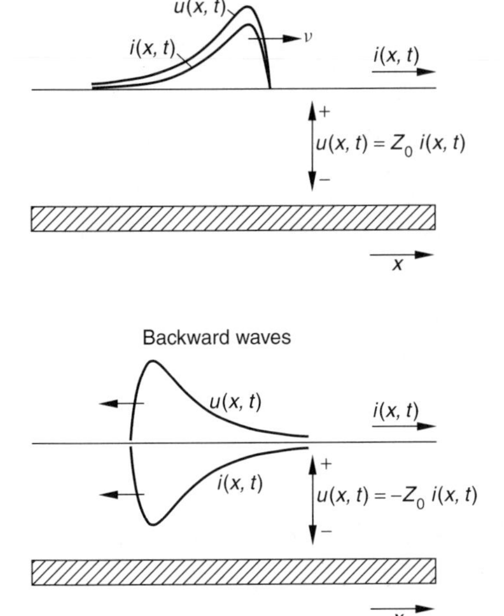


**Figure 3.5** Forward- and backward-travelling waves and their polarity

$f_2[t - (x/v)]$ is a function describing a wave propagating in the positive $x$-direction, called *the forward wave* (see Figure 3.5). The minus sign for the backward current wave related to the backward voltage wave accounts for the fact that the backward energy wave, when the Poynting vector $\boldsymbol{P} = \boldsymbol{E} \times \boldsymbol{H}$ is calculated, has a negative sign because it propagates in the negative $x$-direction.

### 3.4.2 The Distortionless Line

Typical of power systems is that the overhead lines and underground cables have small losses. This means that $R/L$ and $G/C$ are small and that the attenuation constant $\alpha$ and the phase constant $\beta$ are small compared with the rate of change $p = \partial/\partial t$ of the voltage and current waves. For a distortionless line, $R/L = G/C$, which means that the losses of the electric field equal the losses of the magnetic field. As is the case with the lossless line, the voltage and current waves carry an equal amount of energy to every point on the line, and the rate at which they lose energy at every point on the line is also equal. There is, however, no need for an exchange of energy between the voltage and current wave; therefore, the voltage and current waves are undistorted, keep their original shape, and are only attenuated in amplitude. This is the case for power systems, and the voltage wave equation (Equation (3.34)) reduces to

$$u(x,t) = \mathrm{e}^{\alpha x/v} f_1\left(t + \frac{x}{v}\right) + \mathrm{e}^{-\alpha x/v} f_2\left(t - \frac{x}{v}\right) \tag{3.44}$$

and the current wave equation (Equation (3.44)) can be written as

$$i(x,t) = -\frac{1}{Z_0}\left[\mathrm{e}^{\alpha x/v} f_1\left(t + \frac{x}{v}\right) - \mathrm{e}^{-\alpha x/v} f_2\left(t - \frac{x}{v}\right)\right] \tag{3.45}$$

Thus, except for an attenuation constant, the characteristics of a distortionless line, are the same as that of a lossless line: a constant wave velocity $v = 1/(LC)^{1/2}$, and a constant real characteristic impedance $Z_0 = (L/C)^{1/2}$.

## *3.5 REFLECTION AND REFRACTION OF TRAVELLING WAVES*

When an electromagnetic wave propagates along a transmission line with a certain characteristic impedance, there is a fixed relation between the voltage and current waves. But what happens if the wave arrives at a discontinuity, such as an open circuit or a short-circuit, or at a point on the line where the characteristic impedance $Z = (L/C)^{1/2}$ changes, for instance, when an overhead transmission line is terminated with a cable or a transformer? Because of the mismatch in characteristic impedance, an adjustment of the voltage and current waves must occur. At the discontinuity, a part of the energy is let through and a part of the energy is reflected and travels back. At the discontinuity, the voltage and current waves are continuous. In addition, the total amount of energy in the electromagnetic wave remains constant, if losses are neglected. Figure 3.6 shows the case in which an overhead transmission line is terminated with an underground cable.

When, for the sake of simplicity, both the overhead line and the underground cable are assumed to be without losses, $R = 0$ and $G = 0$, then the expressions for the characteristic impedances are

$$Z_L = \sqrt{\frac{L_L}{C_L}}$$

and

$$Z_C = \sqrt{\frac{L_C}{C_C}} \tag{3.46}$$

The forward wave is called the *incident wave* and it travels on the overhead line toward the cable. The positive $x$-direction in Figure 3.6 is from left

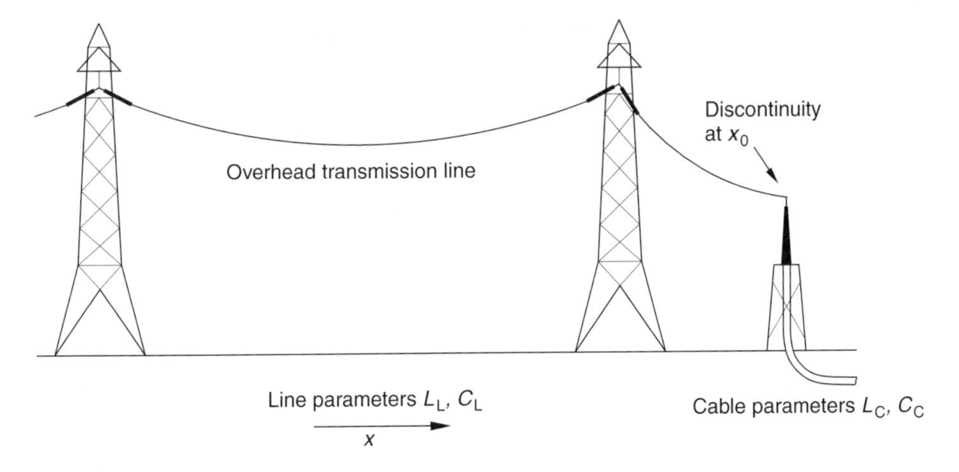


**Figure 3.6** Overhead transmission line terminated with an underground cable

to right and the line-cable joint is at $x_0$. The incident wave reaches the discontinuity $x_0$ at $t = t_0$. In our equations, the incident waves have subscript 1, the reflected waves have subscript 2, and the waves that are let through have subscript 3. Voltage and current waves are continuous at the line–cable joint:

$$v_L(x_0, t) = v_C(x_0, t)$$

and

$$i_L(x_0, t) = i_C(x_0, t) \tag{3.47}$$

This leads to the conclusion that for $t > t_0$, at the discontinuity at $x_0$, the incident waves and the reflected waves equal the waves that are let through. For the voltage waves, we can write

$$v_{1L}(x_0, t) + v_{2L}(x_0, t) = v_{3C}(x_0, t) \tag{3.48}$$

The reflected current wave has a negative sign:

$$\frac{v_{1L}(x_0, t)}{Z_L} - \frac{v_{2L}(x_0, t)}{Z_L} = \frac{v_{3C}(x_0, t)}{Z_C} \tag{3.49}$$

From Equation (3.48) and Equation (3.49), for the voltage and current waves that are let through

$$v_{3C}(x_0, t) = \left\{\frac{2Z_C}{Z_L + Z_C}\right\} v_{1L}(x_0, t) \tag{3.50}$$

$$i_{3C}(x_0, t) = \left\{\frac{2Z_L}{Z_L + Z_C}\right\} i_{1L}(x_0, t) \tag{3.51}$$

and for the reflected waves

$$v_{2L}(x_0, t) = \left\{\frac{Z_C - Z_L}{Z_L + Z_C}\right\} v_{1L}(x_0, t) \tag{3.52}$$

$$i_{2L}(x_0, t) = \left\{\frac{Z_L - Z_C}{Z_L + Z_C}\right\} i_{1L}(x_0, t) \tag{3.53}$$

For a transition from overhead line to cable in a substation, an incoming voltage wave is reduced in amplitude. When an incoming overhead line is connected to a substation by means of a cable of short length, the cable protects the substation against overvoltages. When lightning strikes the substation, the opposite occurs – a magnified voltage wave leaves the substation and travels over the overhead line.

When the incident wave encounters a short-circuit at $x_0$, which means, in our example, that $Z_C = 0$, the voltage wave disappears at $x_0$ and the current wave doubles in amplitude; the wave energy is stored in the magnetic field only. When the incident wave encounters an open circuit at $x_0$, the voltage wave is doubled in amplitude and the incident current wave is nullified by the reflected current wave; the wave energy is stored in the electric field only. The voltage doubling that occurs when a voltage source is switched on an unloaded cable or an unloaded overhead line is called the *Ferranti-effect*, named after the British scientist and engineer Sebastiano Ziani de Ferranti (1864–1930).

When a transmission line is terminated with a load impedance $Z_L$, different from the lines characteristic impedance $Z_0$, the ratio of the complex amplitudes of the reflected and incident voltage waves at the load is called the *voltage reflection coefficient*

$$r = \frac{Z_L - Z_0}{Z_L + Z_0} \tag{3.54}$$

The voltage reflection coefficient $r$ is, in general, a complex quantity with a magnitude smaller than one.

## 3.6 REFLECTION OF TRAVELLING WAVES AGAINST TRANSFORMER AND GENERATOR WINDINGS

When the travelling waves reach windings of transformers and generators, they encounter windings whose physical dimensions are rather small compared with the dimensions of the power system itself. The travelling waves generate oscillations and resonance in the windings and cause standing waves. Because a standing wave results from the superposition of two waves travelling in opposite directions, the voltages and currents in the generator windings and transformer windings should be treated as travelling waves. This implicates that we have to know their characteristic impedance, if we want to calculate the reflection and refraction of voltage and current waves against transformers and generators.

To understand the characteristic impedance of a power transformer, we assume the transformer to consist of a coil mounted on an iron core, as shown in Figure 3.7. The height of the cylindrical winding is

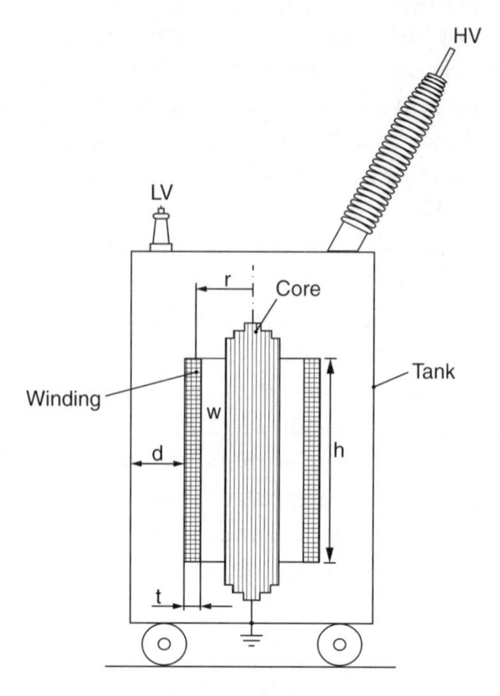


**Figure 3.7** Dimensions of a simplified power transformer $h \gg 2r$, $r \gg w$, $w \gg t$

large compared with its diameter ($h \gg 2r$); the radius of the winding is large compared to the gap $w$ between winding and core ($r \gg w$); and the gap $w$ between winding and core on itself is large compared with the thickness $t$ of the winding($w \gg t$). The distance between tank and winding is $d$, and both the tank and transformer core are assumed to be ideal for conducting. The core and the transformer tank are grounded. The boundary conditions provide constraints on the components of the field vectors as they transit across the boundary between two media. The tangential components of the electric field intensity vector $\boldsymbol{E}$ and the magnetic field intensity vector $\boldsymbol{H}$ must be continuous across the boundary between two physical media – medium 1 and medium 2

$$\boldsymbol{E}_{t1} = \boldsymbol{E}_{t2} \quad \text{and} \quad \boldsymbol{H}_{t1} = \boldsymbol{H}_{t2} \tag{3.55}$$

In addition, the normal components of the electric flux density vector $\boldsymbol{D}$ and the magnetic flux density vector $\boldsymbol{B}$ must be continuous across the boundary between two physical media

$$\boldsymbol{D}_{n1} = \boldsymbol{D}_{n2} \quad \text{and} \quad \boldsymbol{B}_{n1} = \mathrm{B}_{n2} \tag{3.56}$$

The core and the tank wall are assumed to be perfectly conductive; this is characterised by an infinite conductivity $\sigma = \infty$. The infinite conductivity makes all the fields in the perfect conductor to be zero. This requires that the tangential component of $E_1$ of medium 1 adjacent to medium 2 (the perfect conductor) must be zero at the boundary. In addition, the normal component of $\boldsymbol{B}_1$ must be zero at the boundary. The tangential component of $\boldsymbol{H}_1$ and the normal component of $\boldsymbol{D}_1$ cannot be zero; otherwise, the resulting fields in medium 1 would be overspecified. The tangential component of the magnetic field in medium 1, $\boldsymbol{H}_1$, equals the surface current density on the interface, and the normal component of the electric flux $\boldsymbol{D}_1$ equals the surface charge density on the interface (Figure 3.8). This simplifies the mathematical calculations. When the right contour and surface is chosen, the capacitance and inductance of the winding can be calculated and this result can be used to calculate the characteristic impedance.

When a constant voltage $V$ is present between the winding terminal and the ground, there is a linear voltage profile along the coil that results in an inner electric field $E_{\mathrm{inner}}$ between coil and core and an outer electric field $E_{\mathrm{outer}}$ between coil and tank. Gauss' law for the electric field says that

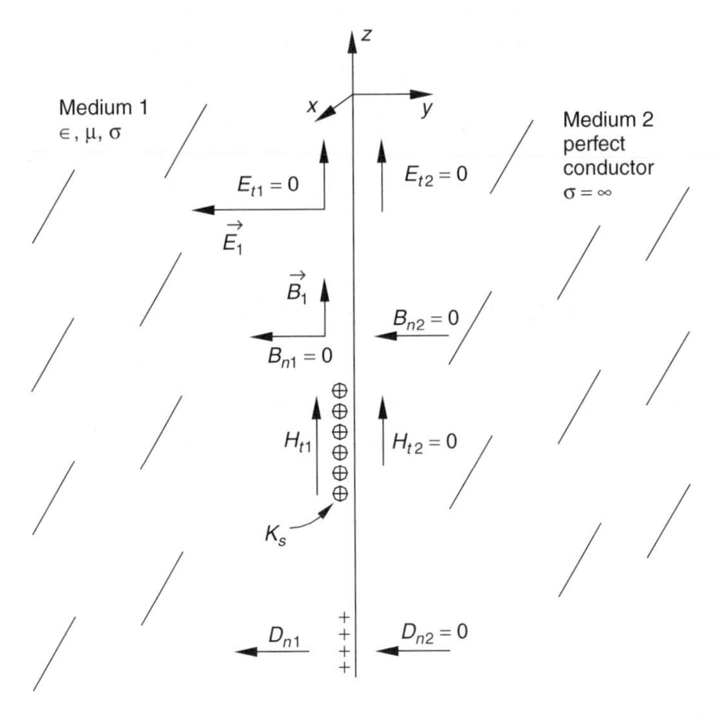


**Figure 3.8** Illustration of the boundary conditions in which one medium is a perfect conductor

*the net flux of the electric flux density vector out of a closed surface S is equivalent to the net positive charge enclosed by that surface.* Written in integral form

$$\iint_S D_n \cdot \mathrm{d}S = \rho \tag{3.57}$$

The total charge $\rho$ can thus be calculated with Gauss' law

$$\rho = 2\pi rh\varepsilon(E_{\text{outer}} + E_{\text{inner}}) = 2\pi rh\varepsilon\left(\frac{V}{d} + \frac{V}{w}\right) \tag{3.58}$$

In this expression, $\varepsilon$ is the permittivity of the medium between coil and core and the medium between coil and tank. The total capacitance to ground is

$$C = \frac{\rho}{V} = 2\pi rh\varepsilon\left(\frac{1}{d} + \frac{1}{w}\right) = 2\pi rh\varepsilon\left(\frac{w+d}{wd}\right) \tag{3.59}$$

When a constant current $i$ flows through the perfectly conductive coil, the surface current density exists on the inside and outside of the coil, which is orthogonal to the tangential component of the magnetic field vector $\boldsymbol{H}$ in the medium between the coil and the core and the medium between the coil and the tank. The magnetic flux inside the coil and outside the coil are equal in magnitude but have an opposite sign. The total magnetic flux in the medium between the coil and the core and the coil and the tank can be calculated with Ampere's law

$$\oint_C H \cdot \mathrm{d}l = \iint_S J \cdot \mathrm{d}S + \frac{\mathrm{d}}{\mathrm{d}t}\iint_S D \cdot \mathrm{d}S \tag{3.60}$$

Because we assume that the current does not change in time, the displacement current is zero, and the magnetic flux is determined by the *total conduction current that penetrates the surface S bounded by the contour C*. When $N$ is the number of windings of the coil, the total magnetic flux $\Phi$ is

$$\Phi = \frac{4\pi\mu Ni2\pi r}{\left(\frac{h}{w} + \frac{h}{d}\right)} \tag{3.61}$$

In this expression, $\mu$ is the permeability of the medium between coil and core and of the medium between coil and tank. The total inductance of the coil is

$$L = \frac{N\Phi}{i} = 4\pi\mu\left(\frac{2\pi r}{h}\right)N^2\left(\frac{wd}{w+d}\right) \tag{3.62}$$

The expression for the characteristic impedance is

$$Z = \sqrt{\frac{L}{C}} = 2N\left(\frac{wd}{hw+hd}\right)\sqrt{\frac{\mu_0}{\varepsilon_0}}\sqrt{\frac{\mu_r}{\varepsilon_r}} \tag{3.63}$$

We learn from this expression that the characteristic impedance of a single transformer or of a generator coil is determined by the permeability and the permittivity of the medium, the distance $w$ between the coil and the core, and the number of windings per unit length $N/h$.

A typical value of the characteristic impedance for the low-voltage windings of a 500-kVA distribution transformer is 600 Ω and for the high-voltage windings it is 5000 Ω. For a 50-MVA power transformer, typical values are 100 Ω for the low-voltage windings and 1000 Ω for the high-voltage windings.

## 3.7 *THE ORIGIN OF TRANSIENT RECOVERY VOLTAGES*

After interruption of a short-circuit current by a high-voltage fuse or a power circuit breaker, a *transient recovery voltage* (TRV) appears across the terminals of the interrupting device.

When the short-circuit current is interrupted at current zero, there is still magnetic energy stored in the leakage inductance of the transformers in the substation, in the self-inductance of the stator and field windings of the supplying generators, and in the inductance of the connected bus bars, the overhead lines and the underground cables.

Electromagnetic waves propagate through the system even after current interruption, which is caused by the sudden change in the configuration of the system. These voltage waves reflect against transformers, increase in amplitude, and travel back to the terminals of the interrupting device. Let us have a look at an outdoor substation layout, as depicted in Figure 3.9.

The circuit breaker CB interrupts an ungrounded three-phase short circuit on an overhead transmission line close to the breaker terminals, a so-called *bolted terminal fault*. In Figure 3.10, the three-phase representation is shown.

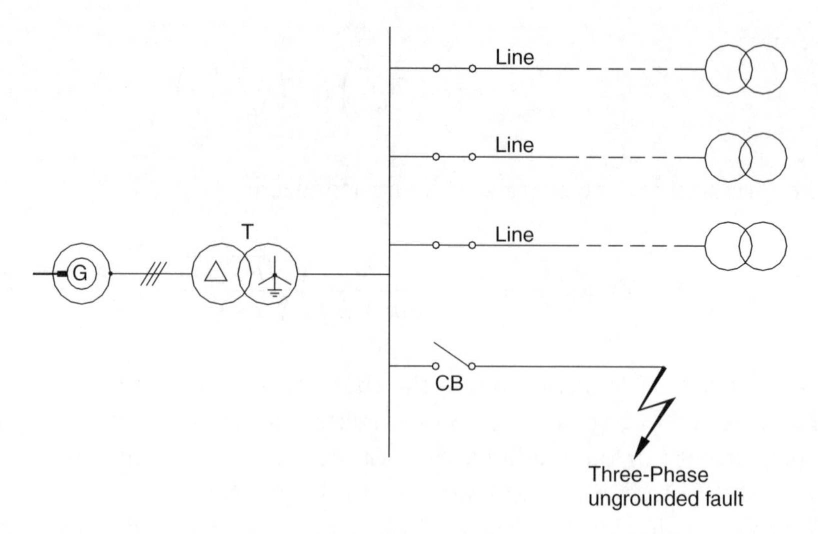


**Figure 3.9** One-line diagram of an outdoor substation with a three-phase ungrounded fault on a feeder

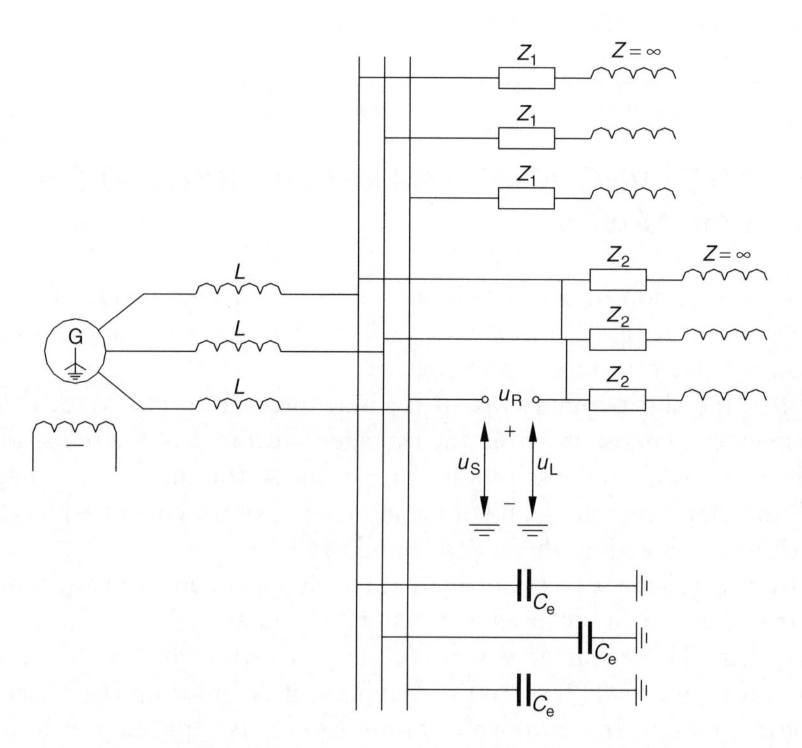


**Figure 3.10** Three-phase representation of an outdoor substation with a three-phase ungrounded fault on a feeder

The supplying generators and transformers are modelled as a three-phase voltage source and a series inductance. The natural capacitance of the transformer high-voltage windings and bushings, the bus bars, and the voltage and current transformers are represented by a lumped capacitance $C_e$. The adjacent feeders and the short-circuited feeder are modelled by their characteristic impedance. For the high-voltage circuit breaker interrupting the three-phase ungrounded fault, we assume that phase $R$ is the first phase to clear and the other poles $S$ and $T$ of the breaker are still arcing. Because the three-phase fault is isolated and the neutral of the supply is solidly grounded, the first-pole-to-clear-factor (FTPC) is 1.5 (see Chapter 2, *Transient Analysis of Three-Phase Power Systems*). The TRV across the terminals of the interrupting device is the difference between the TRV generated by the network elements at the supply side $u_S$ and the TRV generated at the load side $u_L$.

When we examine the three-phase representation of the network of Figure 3.10 from the breaker terminals, we can draw up a simplified network representation as depicted in Figure 3.11. For large substations, the number of adjacent feeders $n$ in parallel results in a characteristic impedance that is considerably smaller than the characteristic impedance of the faulted line $Z_2 \gg Z_1$.

If we also neglect the natural capacitance $C_e$, as is seen from the breaker terminals, the network can be further simplified to a parallel connection of a resistance (for a lossless line, the characteristic impedance is a real number) and an inductance as shown in Figure 3.12.

To find the response of the parallel network of Figure 3.12 on the interruption of a short-circuit current, we use the principle of superposition and inject a current $i = \sqrt{2}\, I\omega t$, which is an approximation of the first part of the interrupted fault current. The resulting transient response across

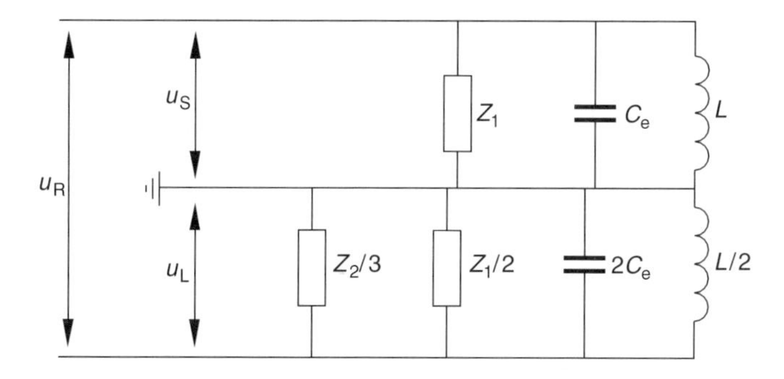


**Figure 3.11** Simplified network representation seen from the breaker terminals

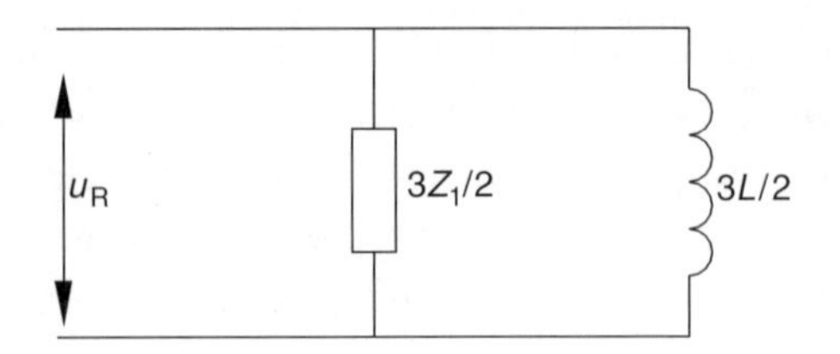


Figure 3.12 Simplification of the network of Figure 3.11

the terminals of the interrupting device (see Figure 3.13) is

$$u_R(t) = \frac{3}{2} L\sqrt{2}I\omega \left(1 - e^{-Z_1 t/L}\right) \tag{3.64}$$

Because we started the analysis from the three-phase representation, the FTPC factor is automatically taken into account. The rate of rise of the TRV at current zero is determined by the value of the characteristic impedance $Z_1$. When the natural capacitance $C_e$ is not neglected, this capacitance causes a time delay $t_d = C_e Z_1$ at current zero. For large values of the capacitance $C_e$, the TRV is the response of a parallel RLC network and the waveform has a (1-cosine) shape.

So far, we have considered the voltage response at the instant of current interruption and the first few microseconds thereafter. In this time period, the contribution to the TRV is mainly from oscillating lumped elements. After the first few microseconds, reflected travelling waves arrive at the terminals of the interrupting device and these travelling waves contribute to the transient recovery waveform (Figure 3.14). It is this cumulation of reflected electromagnetic waves combined with local oscillations that gives the TRV its irregular waveform. The shape of the TRV waveform depends on the current that was interrupted, the line and cable lengths, the propagation velocity of the electromagnetic waves, and the reflection rates at the discontinuities.

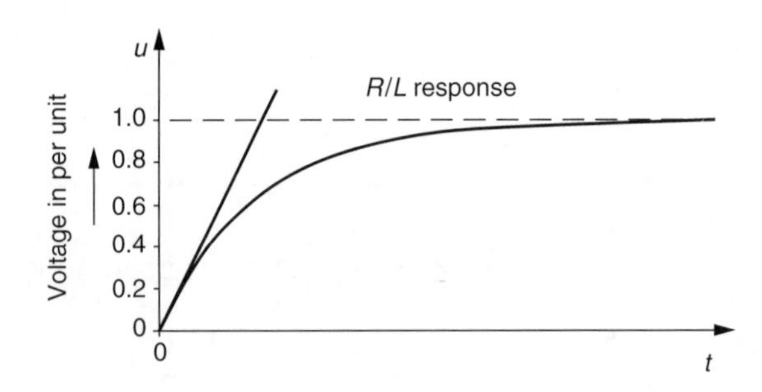


Figure 3.13 First-approximation of the TRV across the breaker terminals

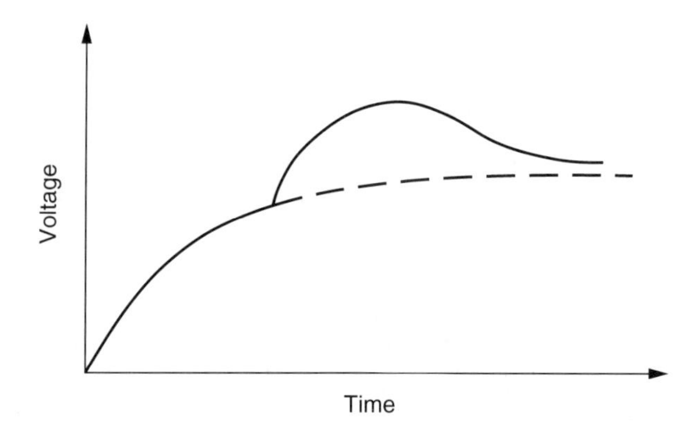


**Figure 3.14** Generalised shape of the TRV formed by local oscillations and reflected travelling waves

## *3.8 THE LATTICE DIAGRAM*

Lattice diagrams have been introduced by Bewley and are of great help while making calculations with travelling waves. To demonstrate the application of lattice diagrams, we will investigate what happens when in Figure 3.15 the circuit breaker closes and switches an energised transformer on an unloaded overhead line in series with an unloaded cable connected to an unloaded transformer. When the circuit breaker closes at $t = 0$, the voltage of the supplying transformer is at its maximum.

The following characteristics apply:

Overhead line: Characteristic impedance $Z_{line} = 400\ \Omega$
Line length $L_{line} = 3000\ \text{m}$
Propagation velocity of the electromagnetic waves $v_{line} = 300\,000\ \text{km/s}$
Travel time $\tau_{line} = 10\ \mu\text{s}$

Cable Characteristic impedance $Z_{cable} = 40\ \Omega$
Cable length $L_{cable} = 100\ \text{m}$

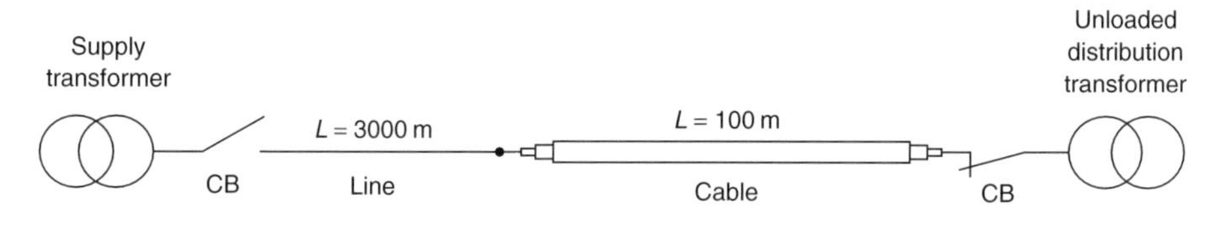


**Figure 3.15** One-line diagram of a circuit breaker closing on a series connection of an unloaded overhead line, an unloaded cable, and an unloaded distribution transformer

Propagation velocity of the electromagnetic waves
$\nu_{\text{cable}} = 100\,000$ km/s
Travel time $\tau_{\text{cable}} = 1$ μs

Source voltage: $v(t) = \cos(\omega t)$ per unit, frequency 50 Hz

When the circuit breaker closes at $t = 0$, the supply voltage has its maximum value of one per unit and a voltage wave, with an amplitude of one per unit, travels along the overhead line. As the voltage wave reaches the discontinuity where the overhead line is connected to the cable, the incident wave breaks up into a reflected wave with an amplitude of $(40 - 400)/(40 + 400) = -0.818$ per unit and a transmitted wave with an amplitude of $2 \times 40/(40 + 400) = 0.182$ per unit. The transmitted

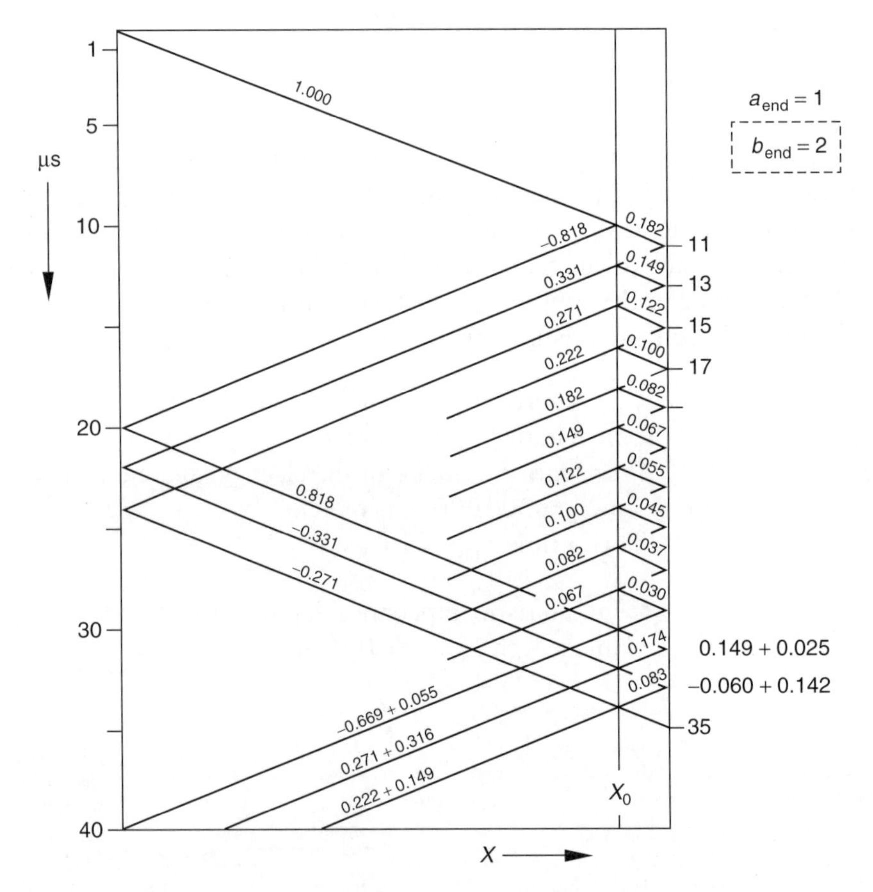


**Figure 3.16** Lattice diagram showing the reflections and the refractions of the travelling waves of the circuit of Figure 3.15. The overhead line is terminated with an underground cable at $X_0$

wave propagates along the cable toward the unloaded transformer. The unloaded transformer has a very high characteristic impedance compared to the cable and can thus be treated as an open end; the voltage is doubled because the reflected wave adds up to the incident voltage wave. The voltage wave travelling from the unloaded transformer back along the cable on its turn breaks up at the discontinuity where the overhead line and the cable are connected into a reflected wave that bounces back to the unloaded transformer, and a transmitted wave travelling to the source.

This process goes on and on. Because the wave phenomena play their game on the microseconds timescale, the source voltage can be assumed to remain constant during the process of reflection and refraction of the electromagnetic waves. If we make a diagram and put the travel time along the vertical axis and the distance along the horizontal axis, we get the so-called lattice diagram (Figure 3.16). In this lattice diagram, we do the bookkeeping of the reflections and the breaking of the travelling waves. When we start with an incident wave of one per unit, the refraction and reflection constants at the discontinuities for the voltage waves (Equation (3.50) and Equation (3.52)) are also placed along the horizontal axis to determine the amplitude of the voltage wave at a certain place along the line and the cable, being the superposition of the amplitudes of the waves having arrived at that place at that instant.

The amplitudes of the voltages at the supply transformer $U_{\text{start}}$, the place where the line and the cable are connected $U_{\text{dis}}$, and at the terminals of the

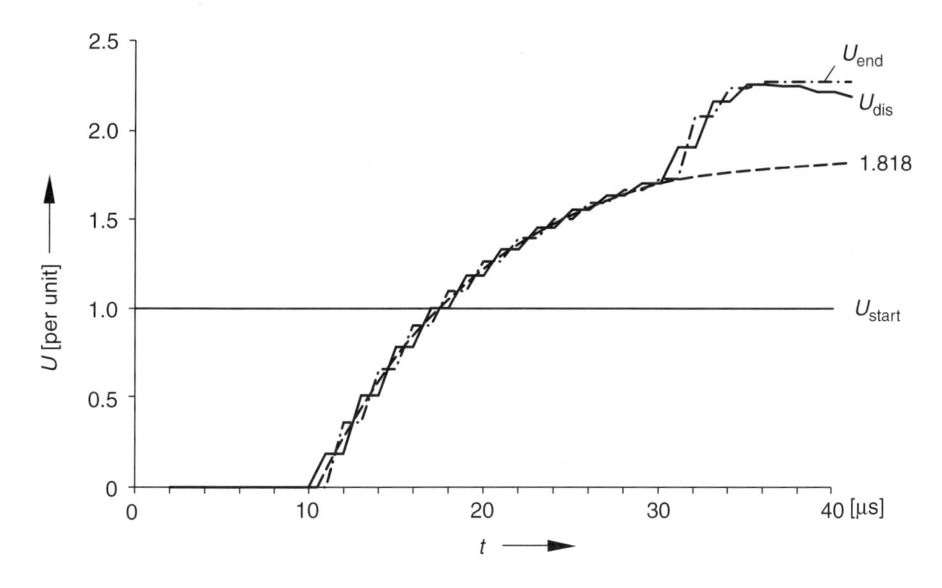


**Figure 3.17** Shape of the TRV as an addition of the reflections of the travelling waves

unloaded transformer $U_{end}$ can be constructed with the help of the lattice diagram of Figure 3.16, and the voltage curves are drawn in Figure 3.17.

We see from Figure 3.17 that the voltages are discontinuous in time. In the shape of the voltage at the receiving end, we recognise the response of a lumped RC network. This is logical because the distributed capacitance of the cable is charged and this takes some time. When the circuit breaker closes, the voltage at the terminals of the unloaded transformer reaches its steady-state value gradually until the cable is charged.

## 3.9 REFERENCES FOR FURTHER READING

Baatz, H., *Ueberspannungen* in *Energieversorgungsnetzen*, Part C, Chapters 1–5, Springer-Verlag, Berlin, 1956.

Bewley, L. V., *Travelling Waves on Transmission Systems*, 2nd ed., Chapters 1–4, Dover Publications, Mineola, New York, 1963.

Browne, T. E., Jr., *Circuit Interruption*, Chapter 3, Marcel Dekker, New York, 1984.

Feynman, R. P., *The Feynman Lectures on Physics*, Vol. II Chapter 22, Addison-Wesley, Reading, Massachusetts, 1997, 6th printing.

Garzon, R. D., *High-Voltage Circuit Breakers*, Chapter 3, Marcel Dekker, New York, 1997.

Grainger, J. J. and Stevenson, Jr., W. D., *Power System Analysis*, Chapters 4, 5, McGraw-Hill, New York, 1994.

Greenwood, A., *Electrical Transients in Power Systems*, 2nd ed., Chapter 9, Wiley & Sons, New York, 1991.

Happoldt, H. and Oeding, D., *Elektrische Kraftwerke und Netze*, 5th ed., Chapter 17, Springer-Verlag, Berlin, 1978.

Paul, C. R., *Introduction to Electromagnetic Compatibility*, Chapter 3, Wiley & Sons, New York, 1992.

Ruedenberg, R., *Elektrische Wanderwellen*, 4th ed., Parts I, II, IV, Chapters 1, 6, 17, Springer-Verlag, Berlin, 1962; Ruedenberg, R., *Electrical Shock Waves in Power Systems*, Parts I, II, IV, Chapters 1, 6, 17, Harvard University Press, Cambridge, Massachusetts, 1968.

Wagner, C. L. and Smith, H. M., "Analysis of transient recovery voltage (TRV) rating concepts," *IEEE T-PAS*, **103**(11), 3354–3362, 1984.

# 4

# Circuit Breakers

For the analysis of simple switching transients and for carrying out large system studies, it is often sufficient to model a circuit breaker as an ideal switch. When studying arc–circuit interaction, wherein, the influence of the electric arc on the system elements is of importance, a thorough knowledge about the physical processes between the circuit breaker contacts is absolutely necessary.

A high-voltage circuit breaker is an indispensable piece of equipment in the power system. The main task of a circuit breaker is to interrupt fault currents and to isolate faulted parts of the system. Besides short-circuit currents, a circuit breaker must also be able to interrupt a wide variety of other currents at system voltage such as capacitive currents, small inductive currents, and load currents. We require the following from a circuit breaker:

- In closed position it is a good conductor;
- In open position it behaves as a good isolator between system parts;
- It changes in a very short period of time from close to open;
- It does not cause overvoltages during switching;
- It is reliable in its operation.

The electric arc is, except from power semiconductors, the only known element that is able to change from a conducting to a nonconducting state in a short period of time. In high-voltage circuit breakers, the electric arc is a high-pressure arc burning in oil, air, or sulphur hexafluoride ($SF_6$). In medium-voltage breakers more often, the low-pressure arc burning in vacuum is applied to interrupt the current. The current interruption is performed by cooling the arc plasma so that the electric arc, which is formed between the breaker contacts after contact separation, disappears. This cooling process or arc-extinguishing can be done in different ways. Power circuit breakers are categorised according to the extinguishing medium in the interrupting chamber in which the arc is formed. That is the reason why we speak of oil, air-blast, $SF_6$, and vacuum circuit breakers.

In 1907, the first oil circuit breaker was patented by J. N. Kelman in the United States. The equipment was hardly more than a pair of contacts submersed in a tank filled with oil. It was the time of discovery by experiments and most of the breaker design was done by trial and error in the power system itself. In 1956, the basic patent on circuit breakers employing $SF_6$ was issued to T. E. Browne, F. J. Lingal, and A. P. Strom. Presently the majority of the high-voltage circuit breakers use $SF_6$ as extinguishing medium.

J. Slepian has done much to clarify the nature of the circuit breaker problem, because the electric arc proved to be a highly intractable and complex phenomenon. Each new refinement in experimental technique threw up more theoretical problems. The practical development of circuit breakers was, especially in the beginning, somewhat pragmatic, and design was rarely possible as deduction from scientific principles. A lot of development testing was necessary in the high-power laboratory. A great step forward in understanding arc–circuit interaction was made in 1939 when A. M. Cassie published the paper with his well-known equation for the dynamics of the arc and then in 1943 O. Mayr followed with the supplement that takes care of the time interval around current zero. Much work was done afterwards to refine the mathematics of those equations and to confirm their physical validity through practical measurements. It becomes clear that current interruption by an electrical arc is a complex physical process when we realise that the interruption process takes place in microseconds, the plasma temperature in the high-current region is more the 10 000 K, and the temperature decay around current zero is about 2000 K/μs per microsecond while the gas movements are supersonic.

Until recently, scientists have succeeded in designing a new high-voltage circuit breaker on the drawing table only; testing in the high-power laboratory still remains necessary. Yet, the understanding of the current interruption process has led to $SF_6$ circuit breakers capable of interrupting 63 kA at 550 kV with a single interrupting element.

## 4.1 THE SWITCHING ARC

The electric arc in a circuit breaker plays the key role in the interruption process and is therefore often addressed as *switching arc*. The electric arc is a plasma channel between the breaker contacts formed after a gas discharge in the extinguishing medium. When a current flows through a circuit breaker and the contacts of the breaker part, driven by the mechanism, the magnetic energy stored in the inductances of the power system forces the current to flow. Just before contact separation, the breaker contacts touch each other at a very small surface area and the resulting high current density makes the contact material to melt. The melting contact material virtually explodes and this leads to a gas discharge in the surrounding medium either air, oil, or $SF_6$.

When the molecular kinetic energy exceeds the combination energy, matter changes from a solid state into a liquid state. When more energy is added by an increase in temperature and the Van der Waals forces are overcome, matter changes from a liquid state into a gaseous state. A further increase in temperature gives the individual molecules so much energy that they dissociate into separate atoms, and if the energy level is increased even further, orbital electrons of the atoms dissociate into free moving electrons, leaving positive ions. This is called the *plasma state*. Because of the free electrons and the heavier positive ions in the high-temperature plasma channel, the plasma channel is highly conducting and the current continues to flow after contact separation.

Nitrogen, the main component of air, dissociates into separate atoms ($N_2 \rightarrow 2N$) at approximately 5000 K and ionises ($N \rightarrow N^+ + e$) above 8000 K. $SF_6$ dissociates into sulphur atoms and fluorine atoms at approximately 1800 K and ionises at temperatures between 5000 and 6000 K. For higher temperatures, the conductivity increases rapidly. The thermal ionisation, as a result of the high temperatures in the electric arc, is caused by collisions between the fast-moving electrons and photons, the slower-moving positively charged ions and the neutral atoms. At the same time, there is also a recombination process when electrons and positively charged ions recombine to a neutral atom. When there is a thermal equilibrium, the rate of ionisation is in balance with the rate of recombination.

The relation between the gas pressure $P$, the temperature $T$, and the fraction of the atoms that is ionised $f$ is given by Saha's equation

$$\frac{f^2}{1-f^2}P = 3.16 * 10^{-7} T^{5/2} \mathrm{e}^{-eV_i/kT} \tag{4.1}$$

with $e = 1.6 * 10^{-19}$, the charge of an electron
$V_i$ = ionisation potential of the gaseous medium
$k = 1.38 * 10^{-23}$, Boltzmann's constant

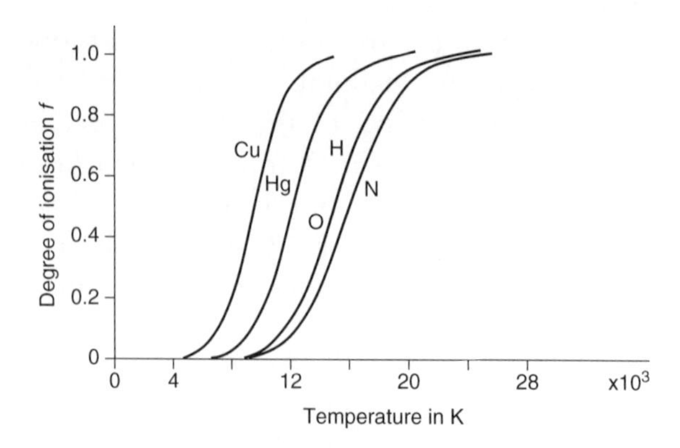


**Figure 4.1** Degree of thermal ionisation for some metal vapours and atomic gases

Saha's relation is shown in graphical form for oxygen, hydrogen, and nitrogen and for the metal vapours of copper and mercury in Figure 4.1.

The graph of Figure 4.1 shows that thermal ionisation can be used to switch between a conducting state ($f = 1$) and an isolating state ($f = 0$). Because of the rather steep slope of the function between the temperature and the degree of ionisation, reduction of the average kinetic energy level of the moving particles by cooling with cold gas is a very good way to bring the arc channel from a conducting to a nonconducting state. Equation (4.1) implies that, because the temperature $T$ cannot change instantaneously, it takes a certain amount of time before thermal equilibrium is reached after changing from the conducting to the non-conducting state. This *conductivity time constant* depends on both the ion–electron recombination speed and the particle-velocity distribution. The time to reach a local molecule-atom velocity equilibrium is in the order of $10^{-8}$ s and the time needed to reach a local electron velocity equilibrium is in the order of $10^{-10}$ s.

The physical mechanisms that play a role in the electron–ion recombination process have time constants in the order of $10^{-7} - 10^{-8}$ s. This means that the time needed to reach ionisation equilibrium is considerably shorter than the rate of change in the electrical phenomena from the power system during the current interruption period. For this reason, the circuit breaker arc can be assumed to be in a thermal–ionisation equilibrium for all electric transient phenomena in the power system.

The plasma channel of the electric arc can be divided into three regions – the column in the middle, the cathode, and anode region (see Figure 4.2).

From the arc channel, the potential gradient and the temperature distribution can be measured. Figure 4.3 shows a typical potential distribution along the arc channel between the breaker contacts.

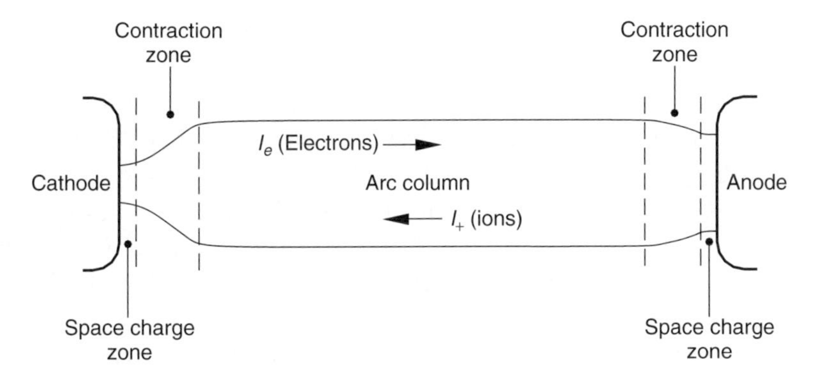


**Figure 4.2** The arc channel can be divided into an arc column, a cathode, and an anode region

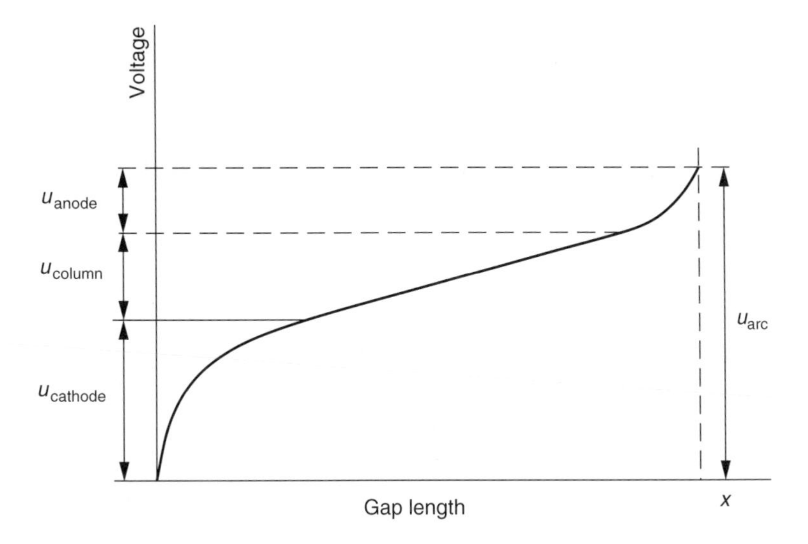


**Figure 4.3** Typical potential distribution along an arc channel

The potential gradient along the arc channel is a function of the arc current, the energy exchange between the plasma channel and the surrounding medium, the pressure, flow velocity, and physical properties of the surrounding medium. The cross-section at a certain point of the cylindrically shaped arc channel adjusts itself such that the potential gradient has the lowest possible value at that point. In the arc column, there are no space charges present, the current flow is maintained by electrons, and there is a balance between the electron charges and the positive ion charges. The peak temperature in the arc column can range from 7000–25 000 K, depending on the arcing medium and configuration of the arcing chamber.

The role of the cathode, surrounded by the cathode region, is to emit the current-carrying electrons into the arc column. A cathode made from *refractory material* with a high boiling point, (e.g. carbon, tungsten, and molybdenum) starts already with the emission of electrons when heated to a temperature below the evaporation temperature – this is called *thermionic emission*. Current densities that can be obtained with this type of cathode are in the order of 10 000 A/cm$^2$. The cooling of the heated cathode spot is relatively slow compared with the rate of change of the transient recovery voltage, which appears across the breaker contacts after the arc has extinguished and the current has been interrupted.

A cathode made from *nonrefractory material* with a low boiling point, such as copper and mercury, experience significant material evaporation. These materials emit electrons at temperatures too low for thermionic emission and the emission of electrons is due to *field emission*. This type of cathode is used in vacuum breakers, in which electrons and ions emanate from individual spots on the cathode surface, each cathode spot supplying 15–150 A, depending on the cathode material. Because of the very small size of the cathode spot, cooling of the heated spot is almost simultaneous with the current decreasing to zero.

The current density in the cathode region is much higher than the current density in the arc column itself. This results in a magnetic field gradient that accelerates the gas flow away from the cathode. This is called the *Maecker effect*.

The role of the anode can be either *passive* or *active*. In its passive mode, the anode serves as a collector of electrons leaving the arc column. In its active mode the anode evaporates, and when this metal vapour is ionised in the anode region, it supplies positive ions to the arc column. Active anodes play a role with vacuum arcs: for high current densities, anode spots are formed and ions contribute to the plasma. This is an undesirable effect because these anode spots do not stop emitting ions at the current zero crossing. Their heat capacity enables the anode spots to evaporate anode material even when the power input is zero and thus can cause the vacuum arc not to extinguish.

Directly after contact separation, when the arc ignites, evaporation of contact material is the main source of charged particles. When the contact distance increases, the evaporation of contact material remains the main source of charged particles for the vacuum arcs. For high-pressure arcs burning in air, oil, or $SF_6$, the effect of evaporation of contact material becomes minimal with increasing contact separation, and the plasma depends mainly on the surrounding medium.

## 4.2 *OIL CIRCUIT BREAKERS*

Circuit breakers built in the beginning of the twentieth century were mainly oil circuit breakers. In those days the breaking capacity of oil circuit breakers was sufficient to meet the required short-circuit level in the substations. Presently, oil and minimum-oil circuit breakers still do their job in various parts of the world but they have left the scene of circuit breaker development.

The first oil circuit breakers were of simple design – an air switch that was put in a tank filled with mineral oil. These oil circuit breakers were of the plain-break type, which means that they were not equipped with any sort of arc quenching device. In 1901, J. N. Kelman of the United States built an oil–water circuit breaker in this way, which is capable of interrupting 200–300 A at 40 kV. Kelman's breaker consisted of two open wooden barrels, each containing a plain-break switch. The two switches were connected in series and operated by one common handle. The wooden barrels contained a mixture of water and oil as extinguishing medium.

In the 1930s, the arcing chamber appeared on stage. The breaker, a metal explosion pot of some form, was fitted with an insulating arcing chamber through which the breaker contacts moved. The arcing chamber, filled with oil, fixes the arc and the increase in pressure inside the arcing chamber improved the cooling effects on the arc considerably. Later, the design of the arcing chamber was further improved by pumping mechanisms, creating a cross flow of oil, giving extra cooling to the arc (Figure 4.4).

A next step in the development of oil circuit breakers was the minimum-oil circuit breaker. The contacts and arcing chamber were placed into

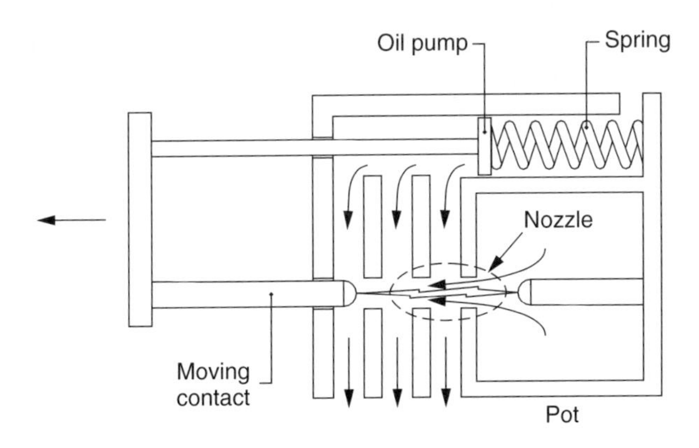


**Figure 4.4** Cross-section of an oil circuit breaker

a porcelain insulator instead of in a bulky metal tank. Bulk-oil circuit breakers with their huge metal tank containing hundreds of litres of mineral oil have been popular in the United States. Minimum-oil circuit breakers conquered the market in Europe.

## *4.3 AIR-BLAST CIRCUIT BREAKERS*

Air is used as insulator in outdoor-type substations and for high-voltage transmission lines. Air can also be used as extinguishing medium for current interruption. At atmospheric pressure, the interrupting capability, however, is limited to low voltage and medium voltage only. For medium-voltage applications up to 50 kV, the breakers are mainly of the magnetic air-blast type in which the arc is blown into a segmented compartment by the magnetic field generated by the fault current. In this way, the arc length, the arc voltage, and the surface of the arc column are increased. The arc voltage decreases the fault current, and the larger arc column surface improves the cooling of the arc channel.

At higher pressure, air has much more cooling power, and air-blast breakers operating with compressed air can interrupt higher currents at considerable higher-voltage levels. Air-blast breakers using compressed air can be of the *axial-blast* or the *cross-blast* type. The cross-blast type air-blast breaker operates similar to the magnetic-type breaker: compressed air blows the arc into a segmented arc-chute compartment. Because the arc voltage increases with the arc length, this is also called *high-resistance interruption*; it has the disadvantage that the energy dissipated during the interruption process is rather high. In the axial-blast design, the arc is cooled in axial direction by the airflow. The current is interrupted when the ionisation level is brought down around current zero. Because the arc voltage hardly increases this is called *low-resistance interruption*.

When operating, air-blast breakers make a lot of noise, especially when the arc is cooled in the free air, as is the case with AEG's free-jet breaker (Freistrahlschalter) design.

## *4.4 $SF_6$ CIRCUIT BREAKERS*

The superior dielectric properties of $SF_6$ were discovered as early as 1920. It lasted until the 1940s before the first development of $SF_6$ circuit breakers began, but it took till 1959 before the first $SF_6$ circuit breaker came to the market. These early designs were descendants of the axial-blast and air-blast circuit breakers, the contacts were mounted inside a tank filled with $SF_6$ gas, and during the current interruption process, the arc was cooled by compressed $SF_6$ gas from a separate reservoir. The liquefying temperature of $SF_6$ gas depends on the pressure but lies in the range of the ambient temperature of the breaker. This means that the $SF_6$ reservoir should be equipped with a heating element that introduces an extra failure possibility for the circuit breaker; when the heating element does not work, the breaker cannot operate.

Therefore the puffer circuit breaker was developed and the so-called *double pressure* breaker disappeared from the market. In the puffer circuit breaker (see Figure 4.5), the opening stroke made by the moving contacts moves a piston, compressing the gas in the puffer chamber and thus causing an axial gas flow along the arc channel. The *nozzle* must be able to withstand the high temperatures without deterioration and is made from teflon.

Presently, the $SF_6$ puffer circuit breaker is the breaker type used for the interruption of the highest short-circuit powers, up to 550 kV–63 kA per interrupter made by Toshiba.

Puffer circuit breakers require a rather strong operating mechanism because the $SF_6$ gas has to be compressed. When interrupting large currents, for instance, in the case of a three-phase fault, the opening speed of the circuit breaker has a tendency to slow down by the thermally generated pressure, and the mechanism (often hydraulic or spring mechanisms) should have enough energy to keep the contacts moving apart. Strong and reliable operating mechanisms are costly and form a substantial part of the price of a breaker. For the lower-voltage range, *self-blast circuit breakers* are now on the market. Self-blast breakers use the thermal energy released by the arc to heat the gas and to increase its pressure. After the

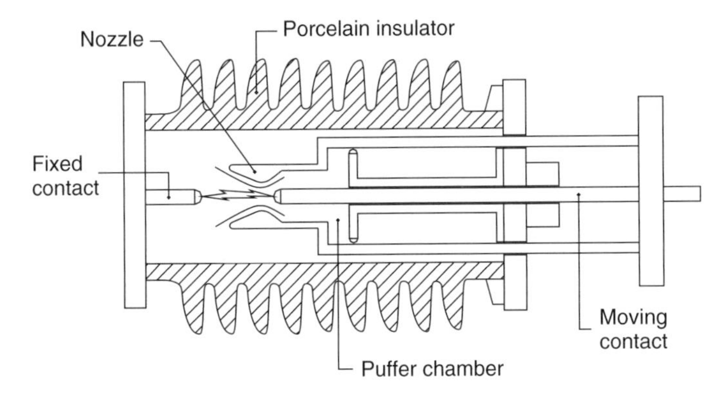


**Figure 4.5** Operating principle of an $SF_6$ puffer circuit breaker

moving contacts are out of the arcing chamber, the heated gas is released along the arc to cool it down. The interruption of small currents can be critical because the developed arcing energy is in that case modest, and sometimes a small puffer is added to assist in the interrupting process. In other designs, a coil carrying the current to be interrupted, creates a magnetic field, which provides a driving force that rotates the arc around the contacts and thus provides additional cooling. This design is called the *rotating-arc circuit breaker*. Both self-blast breakers and rotating arc breakers can be designed with less powerful (and therefore cheaper) mechanisms and are of a more compact design than puffer breakers.

## 4.5 VACUUM CIRCUIT BREAKERS

Between the contacts of a vacuum circuit breaker a vacuum arc takes care of the interruption process. As already discussed in the introduction to this chapter about the switching arc, the vacuum arc differs from the high-pressure arc because the arc burns in vacuum in the absence of an extinguishing medium. The behaviour of the physical processes in the arc column of a vacuum arc is to be understood as a metal surface phenomenon rather than a phenomenon in an insulating medium.

The first experiments with vacuum interrupters took place already in 1926, but it lasted until the 1960s when metallurgical developments made it possible to manufacture gas-free electrodes and when the first

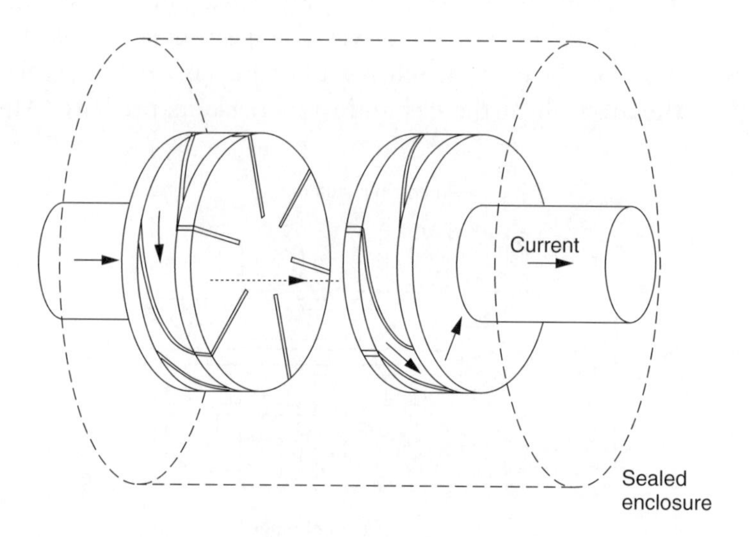


**Figure 4.6** Vacuum interrupter with slits in the contacts to bring the arc in a spiralling motion

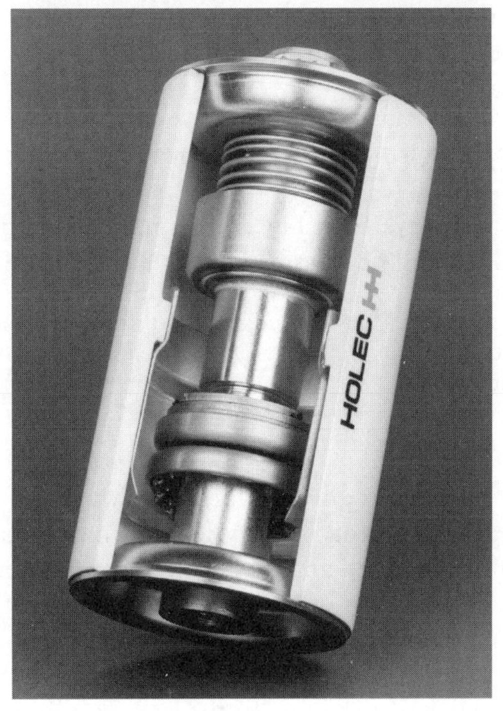


Figure 4.7 The use of horse shoe magnets as is done in the HOLEC interrupters (courtesy of Holec)

practical interrupters were built. There are no mechanical ways to cool the vacuum arc, and the only possibility to influence the arc channel is by means of interaction with a magnetic field. The vacuum arc is the result of a metal-vapour/ion/electron emission phenomenon. To avoid uneven erosion of the surface of the arcing contacts, (especially the surface of the cathode) the arc should be kept *diffused* or in a *spiralling motion.* The latter can be achieved by making slits in the arcing contacts (Figure 4.6) or by applying horseshoe magnets as used in the vacuum interrupters by HOLEC (Figure 4.7).

There is generally less energy required to separate the contacts of a vacuum circuit breaker, and the design of the operating mechanism usually results in reliable and maintenance-free breakers. Vacuum breakers are produced for system voltages up to 72.5 kV and the short-circuit current rating goes up to 31.5 kA.

When the arc current goes to zero it does so in discrete steps of a few amperes to 10 amperes, depending on the contact material. For the last current step to zero, this can cause a noticeable *chopping of the current.* This current chopping in its turn can cause high overvoltages, in particular when the vacuum breaker interrupts a small inductive current, for example, when switching unloaded transformers or stalled motors.

## 4.6 MODELLING OF THE SWITCHING ARC

Arc modelling has always been one of the main topics in circuit breaker research. Arc models can be classified in three categories:

- Black box models (also often called $P$-$\tau$ models)
- Physical models
- Parameter models

The circuit breaker design engineers work mostly with physical arc models when designing a new prototype. Physical arc models are based on the equations of fluid dynamics and obey the laws of thermodynamics in combination with Maxwell's equations. They consist of a large number of differential equations. The arc-plasma is a chemical reaction and, in addition to the conservation of mass equation, describes the rate equations of the different chemical reactions. In the case of a local thermodynamic equilibrium, the rate equations become the equilibrium mass action laws and that, in the simplified case of the reaction of a monatomic gas, becomes the Saha equation, describing the degree of ionisation in the gas. Because the arc-plasma is electrically conducting in the momentum equation also, terms describing the interaction with magnetic fields, either coming from outside or generated by the arc current, are taken into account. Because of the resistive heat dissipation of electric energy calculated with Ohm's law, a volumetric heat generation term is part of the energy equation.

The arc-plasma is strongly radiating and this makes the radiation-transport term in the energy equation very important. A considerable portion of this radiated energy, however, is being reabsorbed in the plasma and this is described by the radiation-transfer equation or by the tabulated value for the net emission coefficient. For every component in the plasma, such as electrons, ions, atoms, and molecular species there is a thermodynamic equation of state present in the equations. When all the physical considerations are taken into account, we arrive at the following set of equations.

Conservation of mass (continuity equation):

$$\frac{\partial \rho}{\partial t} + \mathrm{div}\ (\rho u) = 0 \tag{4.2}$$

Conservation of momentum (Navier–Stokes equation):

$$\rho \frac{\partial u}{\partial t} + \rho(u.\mathrm{grad})u = -\mathrm{grad}(p) \tag{4.3}$$

Conservation of energy

$$\underbrace{\rho \frac{\partial h}{\partial t}}_{\substack{\text{change}\\ \text{of energy}\\ \text{in unit}\\ \text{volume}}} + \underbrace{u.\mathrm{grad}(\rho h)}_{\substack{\text{energy}\\ \text{input by}\\ \text{mass flow}\\ \text{convection}}} - \underbrace{\sigma E^2}_{\substack{\text{Joule}\\ \text{heating}}} = \underbrace{\mathrm{div}\ (\rho u)}_{\substack{\text{work}\\ \text{performed}\\ \text{by flow}}} + \underbrace{\mathrm{div}\ [K.\mathrm{grad}(T)]}_{\substack{\text{thermal}\\ \text{conduction}\\ \text{loss}}} - \underbrace{R[T, \rho]}_{\substack{\text{radiation}\\ \text{loss}}} \tag{4.4}$$

$p$ = pressure
$\rho$ = gas density
$u$ = gas flow velocity
$h$ = enthalpy of gas
$E$ = electric field strength
$\sigma$ = electric conductivity
$K$ = thermal conductivity
$T$ = gas temperature
$R$ = radiation loss
$r$ = arc radius

For one special case the conservation equations become decoupled: a thermally and hydrodynamically fully developed and wall-constricted arc with a negligible axial pressure drop. This case occurs when $z$ is the axial coordinate,

$$\frac{\partial u_z}{\partial z} = 0 \quad u_r = 0 \quad \frac{\partial h}{\partial z} = 0$$

and the energy equation is similar in form to the Elenbaas–Heller equation

$$-\left(\frac{1}{r}\right)\cdot\frac{\partial\left[Kr\left(\frac{\partial T}{\partial r}\right)\right]}{\partial r}+R=\sigma E^2 \tag{4.5}$$

This last equation is very useful when a qualitative evaluation of the effects of current increase or arc-channel-diameter reduction is required.

In *Black box models,* the arc is described by a simple mathematical equation and gives the relation between the arc conductance and measurable parameters such as arc voltage and arc current. These black box models are not suited to design circuit breaker interrupters but are very useful to simulate arc-circuit interaction in network studies. Black box models are based on physical considerations but are, in fact, mathematical models; the behaviour of the arc rather than the physical processes is of importance. Usually, black box models consist of one or two differential equations. *Parameter models* are a variation on black box models in the sense that more complex functions and tables are used for the essential parameters of the black box models.

The classical black box models are the *Cassie model* and the *Mayr model*. Both the Cassie and Mayr equation are a solution of the *general arc equation*.

In its general form, the arc conductance is a function of the power supplied to the plasma channel, the power transported from the plasma channel by cooling and radiation and time:

$$g=F(P_{\text{in}},P_{\text{out}},t)=\frac{i_{\text{arc}}}{u_{\text{arc}}}=\frac{1}{R} \tag{4.6}$$

with

$g$ = the momentary arc conductance
$P_{\text{in}}$ = the power supplied to the plasma channel
$P_{\text{out}}$ = the power transported from the plasma channel
$t$ = time
$i_{\text{arc}}$ = the momentary arc current
$u_{\text{arc}}$ = the momentary arc voltage
$R$ = the momentary arc channel resistance

The momentary arc conductance $g$ varies when $P_{\text{in}}$ and $P_{\text{out}}$ are not in equilibrium. The energy stored in the plasma channel is

$$Q=\int_0^t (P_{\text{in}}-P_{\text{out}})\,\mathrm{d}t \tag{4.7}$$

and the momentary arc conductance can be written as

$$g = F(Q) = F\left[\int_0^t (P_{\text{in}} - P_{\text{out}})\,\mathrm{d}t\right] \tag{4.8}$$

Because we are interested in the change of the arc conductance, Equation (4.8) is differentiated

$$\frac{\mathrm{d}g}{\mathrm{d}t} = \frac{\mathrm{d}F(Q)}{\mathrm{d}Q}\frac{\mathrm{d}Q}{\mathrm{d}t} \tag{4.9}$$

and divided by the momentary arc conductance $g$

$$\frac{1}{g}\frac{\mathrm{d}g}{\mathrm{d}t} = \frac{1}{g}\frac{\mathrm{d}F(Q)}{\mathrm{d}Q}\frac{\mathrm{d}Q}{\mathrm{d}t} \tag{4.10}$$

Differentiation of Equation (4.7) and the result substituted with Equation (4.8) in Equation (4.10) gives us the *general arc equation*:

$$\frac{\mathrm{d}[\ln(g)]}{\mathrm{d}t} = \frac{F'(Q)}{F(Q)}(P_{\text{in}} - P_{\text{out}}) \tag{4.11}$$

To solve this general arc equation, assumptions have to be made. These assumptions, named after their author, give us the different black box models.

In 1939, A. M. Cassie assumed that the arc channel has the shape of a cylinder filled with highly ionised gas with a constant temperature $T$, but with a variable diameter. The heat content per unit of volume remains constant and so does the conductance per unit of volume. Because of the cooling by the gas flow (convection), the arc channel diameter, varies in diameter, but the temperature and the conductance per unit of volume of the remaining plasma channel are not affected.

with

$g_0$ = the conductivity per unit of volume
$P_0$ = the cooling or loss of power per unit of volume
$D$ = the arc channel diameter varying with time
$Q_0$ = the energy content per unit of volume
$u_0$ = $(P_0/g_0)^{1/2}$ the static arc voltage

can for the arc conductance be written as

$$g = F(Q) = Dg_0 \tag{4.12}$$

and for the energy as a function of time

$$Q = DQ_0 \tag{4.13}$$

Combining Equations (4.12) and (4.13) results in

$$g = F(Q) = \frac{Q}{Q_0} g_0 \tag{4.14}$$

$$F'(Q) = \frac{g_0}{Q_0} \tag{4.15}$$

The dissipated power per unit of length

$$P_{\text{out}} = DP_0 = \frac{Q}{Q_0} P_0 \tag{4.16}$$

Equations (4.14), (4.15), and (4.16) substituted in the general arc equation (Equation (4.11)) gives us the *Cassie equation*

$$\frac{\mathrm{d}[\ln(g)]}{\mathrm{d}t} = \frac{P_0}{Q_0}\left(\frac{u_{\text{arc}}^2}{u_0^2} - 1\right) \tag{4.17}$$

The quotient $Q_0/P_0$ is called the arc time constant $\tau$ and can be calculated from the homogeneous differential Equation (4.17)

$$\frac{\mathrm{d}[\ln(g)]}{\mathrm{d}t} = -\frac{P_0}{Q_0} \tag{4.18}$$

A solution satisfying homogeneous Equation (4.18) is

$$g = g_0 \mathrm{e}^{-t/\tau} \tag{4.19}$$

The time constant $\tau$ in Equation (4.19) can be interpreted as the arc time constant parameter with which the arc channel diameter changes.

The *Cassie model* is well suited for studying the behaviour of the arc conductance in the high-current time interval when the plasma temperature is 8000 K or more.

The *Mayr model* describes the arc conductance around current zero. Mayr considered the arc channel to be cylindrical with a constant diameter. The arc column loses its energy by radial heat transport, and the temperature of the arc channel varies more or less exponentially with the temperature and can be expressed as an approximation of Saha's equation (Equation (4.1)):

$$g = F(Q) = ke^{Q/Q_0} \tag{4.20}$$

The cooling or loss of power of the arc channel is assumed to be constant. When this is substituted in the *general arc equation* (4.11), we get

$$\frac{\mathrm{d}[\ln(g)]}{\mathrm{d}t} = \frac{(u_{\mathrm{arc}} i_{\mathrm{arc}} - P_0)}{Q_0} \tag{4.21}$$

At the instant of current zero, the power input $u_{\mathrm{arc}} i_{\mathrm{arc}}$ in the arc channel is zero, and the rate of change of the conductance of the arc channel is

$$\frac{\mathrm{d}g}{\mathrm{d}t} = -g\frac{P_0}{Q_0} \tag{4.22}$$

This is the homogeneous differential equation of Equation (4.21) and has as a solution

$$g = g_0 \mathrm{e}^{-P_0/Q_0 t} \tag{4.23}$$

In this expression $Q_0/P_0$ is the time constant $\tau$ of the arc cooling without thermal input to the arc channel and is called the *arc time constant.*

The *Mayr model* is suited for modelling of the arc in the vicinity of current zero when the temperature of the plasma is below 8000 K.

In 1959, T. E. Browne proposed a composition of the Cassie and Mayr model. The *Browne* model uses a Cassie equation for the high current interval and a Mayr equation for the current zero period.

Other black box models capable of simulating thermal breakdown of the arc channel are the *Avdonin model*, the *Hochrainer model*, the *Kopplin model*, the *Schwarz model*, and the *Urbanek model.* In the 1980s W. R. Rutgers developed the KEMA model based on experiments in KEMA's research laboratory and in KEMA's high-power laboratory. The KEMA model is also capable of simulating a *dielectric breakdown* of the gaseous space between the breaker contacts after current interruption. (See Section 4.7 Arc–circuit interaction) In a hybrid arc model, before current zero, the conductance of the arc channel is calculated with a modified Mayr-equation as proposed by *Haupt*, and in the thermal and dielectric interval around and after current zero, the arc conductance is calculated from the concentration of charged particles and their drift velocity.

## 4.7 ARC–CIRCUIT INTERACTION

Current interruption of a circuit breaker occurs normally at current zero within a time frame of microseconds. In the process of current interruption, several processes take place at the same time. The arc voltage after maintaining a constant value during the high current interval, increases to a peak value, the extinction peak, and then drops to zero with a very steep $du/dt$. The current approaches its zero crossing with a more or less constant $di/dt$ but can be slightly distorted under the influence of the arc voltage. The arc is resistive and therefore the arc voltage and the current reach the zero crossing at the same instant. Around current zero, the energy input in the arc channel is rather low (at current zero there is even no energy input), and when the breaker design is such that the cooling by the extinguishing medium is at its maximum, the current can be interrupted. After current interruption, the still-hot gas between the breaker contacts is stressed by a steep rate of rise of the recovery voltage and in the resulting electric field the present charged particles start to drift and cause a hardly measurable so-called *post-arc current*. The post-arc current, together with the transient recovery voltage, results in energy input in the still-hot gas channel. When the energy input is such that the individual gas molecules dissociate into free electrons and heavier positive ions, the plasma state is created again and current interruption has failed. This is called a *thermal breakdown* of the circuit breaker. When the current interruption is successful, the hot gas channel cools down and the post-arc current disappears; still a *dielectric failure* can occur when the dielectric strength of the gap between the breaker contacts is not sufficient to withstand the transient recovery voltage.

During the current interruption process there is a strong interaction between the physical process between the breaker contacts and the network connected with the terminals of the breaker. The current is determined by the driving voltage and the total series inductance of the network formed by the inductance of the connected lines, cables and bus bars, the synchronous inductance of the generators, and the leakage inductance of the transformers. The transient recovery voltage across the breaker contacts after current interruption is formed by local voltage oscillations and reflected voltage waves. Therefore the simplest lumped-element representation of the power system, as seen from the terminals from the circuit breaker, consists of a voltage source with the value of the system voltage, an inductance with the value of the total short-circuit inductance, a capacitance resembling the stray capacitance of the bus

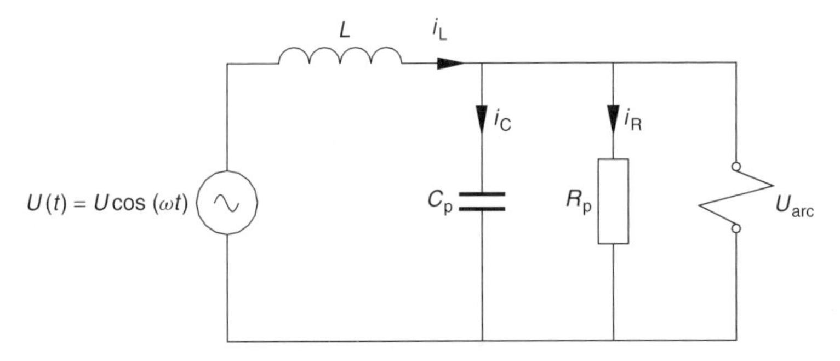


**Figure 4.8** Simple lumped-element representation of the network connected to the breaker terminals

bars, voltage transformers, current transformers, and power transformers in the substation. In parallel with this capacitance, a resistor simulates the characteristic impedance of the connected overhead lines as shown in Figure 4.8.

The $\mathrm{d}i/\mathrm{d}t$ can easily be calculated from the short-circuit current $i(t) = I\sin(\omega t)$:

$$\frac{\mathrm{d}i}{\mathrm{d}t} = \omega I\cos(\omega t) = 2\pi f I_{\mathrm{rms}}\sqrt{2}\cos(\omega t) \tag{4.24}$$

For $t = 0$: $\mathrm{d}i/\mathrm{d}t = 2\pi f I_{\mathrm{rms}}\sqrt{2}$.

As a rule of thumb one can easily remember $\mathrm{d}i/\mathrm{d}t = 0.444\ I_{\mathrm{rms}}$ for 50 Hz, and when $I_{\mathrm{rms}}$ is entered in the kA $\mathrm{d}i/\mathrm{d}t$ is in A/μs.

For testing purposes, the short-circuit currents are divided in so-called duties, each covering a percentage of the full short-circuit current. The IEC 60056 high-voltage circuit breaker standard distinguishes for the breaker terminal fault duties the 10%, 30%, 60% and 100% current ratings. To each duty a specific rate of rise of the recovery voltage is specified and this enables us to calculate the characteristic impedance of the connected network. For instance, for a 100 percent short-circuit current of a 145 kV–40 kA–50 Hz circuit breaker, the specified $\mathrm{d}u/\mathrm{d}t$ is 2000 V/μs and the $\mathrm{d}i/\mathrm{d}t$ is $0.444 * 40 = 17.8$ A/μs. This results in a characteristic impedance of

$$R_{\mathrm{p}} = \frac{\mathrm{d}u/\mathrm{d}t}{\mathrm{d}i/\mathrm{d}t} = \frac{2000}{17.8} = 112\ \Omega \tag{4.25}$$

The capacitance $C_{\mathrm{p}}$ in parallel to the characteristic impedance is rather small: in the nF range. This capacitance $C_{\mathrm{p}}$ causes a delay for the rate of rise of the recovery voltage because immediately after current interruption this capacitance must be charged. The value of this *time delay* is also specified in the IEC 60056, and for a rate of rise of 2000 V/μs connected with the 100 percent duty the delay time is two μs. The value of the time delay equals the $RC$-time of $C_p$ and $R_p$, and in this way the value of the capacitance $C_p$ can be calculated.

The system voltage and the short-circuit current determine the value of the inductance $L$. Figure 4.9 shows the voltage and current traces of the current interruption process around current zero

In the high-current interval, milliseconds before current zero, the $di/dt$ flowing through inductance $L$ is

$$\frac{di_L}{dt} = \frac{(u - u_{arc})}{L} \tag{4.26}$$

and can be assumed to be constant because the arc voltage is constant during the high-current interval.

The constant arc voltage $u_0$ causes a resistive current $i_R = u_0/R_p$ through $R_p$.

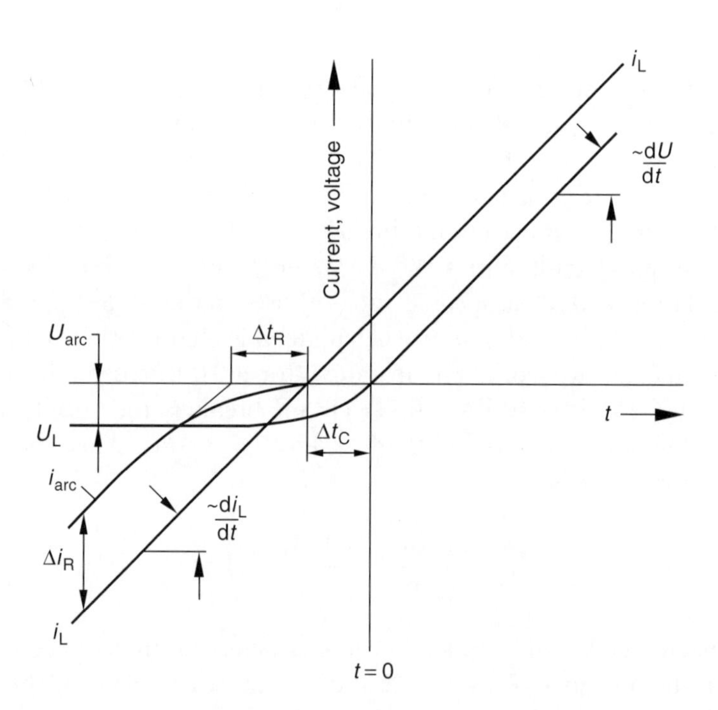


**Figure 4.9** Current and voltage around current zero when a high-voltage circuit breaker interrupts a short-circuit current

$i_R$ leads $i_L$ with a time $\Delta t_R$

$$\Delta t_R = \frac{i_R}{\frac{di_L}{dt}} = \frac{u_0/R_p}{di_L/dt} \tag{4.27}$$

A few microseconds before current zero the arc voltage goes to zero with a steep $du_{arc}/dt$ and causes a capacitive current $i_C$ to flow through $C_p$. This capacitive current $i_C$ lags the current $i_L$ with a time $\Delta t_C$;

$$\Delta t_C = \frac{i_C}{\frac{di_L}{dt}} = \frac{C_p(du_{arc}/dt)}{di_L/dt} \tag{4.28}$$

The presence of the resistor $R_p$ and the capacitance $C_p$ delay the current zero of the linear decaying arc current with a time $\Delta t$

$$\Delta t = \Delta t_R + \Delta t_C = \frac{(u_0/R_p) + C_p du_{arc}/dt}{di_L/dt} \tag{4.29}$$

The parallel resistor, that is, the characteristic impedance of the connected system, and the parallel capacitor achieve commutation of the arc current in the time interval around current zero. The current is shunted, first by $R_p$ and then by $C_p$, the energy input in the arc is reduced, and the cooling mechanism has a longer time period to be effective so that the interruption chance has grown. When the current interruption is successful at current zero the voltage across $C_p$ is equal to zero and the transient recovery voltage building up across the breaker contacts charges first capacitance $C_p$, which causes the so-called *time delay* of the TRV waveform. When a breaker manufacturer wants to increase the interrupting capability of an interrupter design, he can connect a capacitor parallel to the interrupter in the case of an $SF_6$ circuit breaker or a resistor when it concerns an air-blast breaker. For air-blast breakers, a capacitance does not work because of the rather long time constant of physical processes in the arc channel. The disadvantage of using a resistor is that an extra interrupter is needed to switch off the resistor after interruption of the current by the main interrupter.

A high-voltage circuit breaker must be able to interrupt a variety of current values at different power factors. Because of the huge electric power that has to be interrupted in the case of short-circuit currents, it is understandable that fault-current interruption is an important topic.

Specific cases will be analysed in Chapter 5, *Switching Transients*. The interruption of small currents can also lead to situations that are known in literature as *current chopping* and *virtual chopping*. These phenomena are a result of the interaction between the arc and the connected network and lead to high-frequency oscillations through the arc channel, thus forcing a zero crossing before the actual power frequency current zero. When the breaker interrupts this high-frequency current with its steep $\mathrm{d}i/\mathrm{d}t$, the resulting transient recovery voltage has an extremely high peak value and these overvoltages can cause either a re-ignition or, when multiple re-ignitions occur, even a flash-over outside the interrupter.

The difference between *current chopping* and *virtual chopping* is that in the case of *current chopping* the instability of the arc around current zero causes a high-frequency transient current to flow in the neighbouring network elements. This high-frequency current superimposes on the power frequency current of which the amplitude is small and this power frequency current is actually *chopped* to zero. In the case of *virtual chopping* the arc is made unstable through a superimposed high-frequency current caused by oscillations with the neighbouring phases in which current chopping took place. Virtual chopping has been observed for gaseous arcs in air, $SF_6$ and in oil. Especially vacuum arcs are sensitive to current chopping because the vacuum arc consists of cathode spots, and when the arc current approaches zero, current chopping happens in discrete steps of a few Amperes. Virtual chopping, as it occurs for gaseous arcs can be modelled in the following way. Figure 4.10 shows the lumped-element representation of a breaker interrupting a small inductive current.

The small inductive current is supplied by a (inductive) source and, in combination with the mainly inductive load, the current flowing through the breaker is rather small and lags the driving voltage by nearly ninety degrees. For a steady-state arc burning in a gaseous medium, the relation

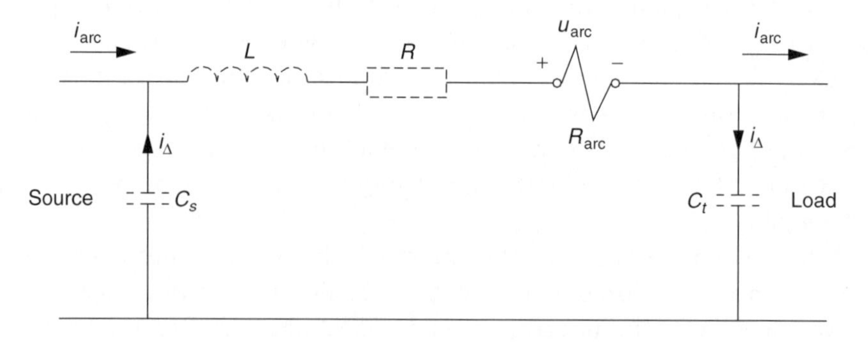


**Figure 4.10** Small inductive current interruption

between the arc voltage and the current is

$$(u_{arc} i_{arc})^{\alpha} = \eta = \text{constant} \tag{4.30}$$

The parameter $\alpha$ has a value between 0.4 and 1.

When a high-frequency oscillation causes a small disturbance $i_{\Delta}$ on the arc current $i_{arc}$ and we assume the arc channel to be in steady-state situation, so that Equation (4.30) gives the relation between the arc current and the arc voltage, applying Kirchhoff's voltage law in Figure 4.10 gives

$$L\frac{d(i_{arc} + i_{\Delta})}{dt} + R(i_{arc} + i_{\Delta}) + R_{arc}(i_{arc} + i_{\Delta}) + \int \frac{i_{\Delta}}{C} dt = u_0 \tag{4.31}$$

with $R_{arc}$ = steady-state arc-resistance, Equation (4.30), $= \dfrac{u_{arc}}{i_{arc}}$

$= \dfrac{\eta}{(i_{arc} + i_{\Delta})^{\alpha+1}}$

$u_0$ = residual charges on both the capacitances $C_s$ and $C_t$

and $C = \dfrac{C_s C_t}{(C_s + C_t)}$

differentiation of Equation (4.31) and substitution of $R_{arc}$ and $dR_{arc}/dt$ gives us

$$L\frac{d^2 i_{\Delta}}{dt^2} + (R - \alpha R_{arc})\frac{di_{\Delta}}{dt} + \frac{i_{\Delta}}{C} = 0 \tag{4.32}$$

The characteristic equation of Equation (4.32) is

$$\lambda^2 + \frac{(R - \alpha R_{arc})}{L}\lambda + \frac{1}{LC} = 0 \tag{4.33}$$

and has the roots

$$\lambda_{1,2} = -\frac{(R - \alpha R_{arc})}{2L} \pm \sqrt{\left(\frac{R - \alpha R_{arc}}{2L}\right)^2 - \frac{1}{LC}} \tag{4.34}$$

The circuit of Figure 4.10 oscillates when $\dfrac{1}{LC} > \left(\dfrac{R - \alpha R_{arc}}{2L}\right)^2$ and when $\alpha R_{arc} > R$ the damping term $\dfrac{R - \alpha R_{arc}}{2L}$ is positive and an induced oscillation can grow in amplitude.

In the case when $R/\alpha < R_{arc} < [R + 2(L/C)^{1/2}]/\alpha$, a small disturbance, when we consider that the physical processes in the arc channel show a stochastic behaviour, can induce an oscillation, which in turn can cause an early zero crossing of the arc current and force the circuit breaker to interrupt the current before the actual power frequency current zero.

The value of the current where chopping occurs, the so-called *chopping current* depends on the extinguishing medium and the value of the capacitance C, which is in the order of 0.01–0.05 μF. For vacuum circuit breakers, the chopping current is only a few Amperes because the chopping phenomenon depends on the contact material. For gaseous arcs, the chopping current can range from a few Amperes for $SF_6$ to several tenths Amperes for air-blast.

## *4.8 REFERENCES FOR FURTHER READING*


Biermanns, J. and Hochrainer, A., "Hochspannungs-Schaltgeräte," *AEG-Mitt.*, Jg. **47**, (H718), S209–S212 (1957).

Browne, Jr., T. E., "A study of arc behaviour near current zero by means of mathematical models," *AIEE Transactions* **67**, 141–153 (1948).

Browne, Jr., T. E., "An approach to mathematical analysis of a-c arc extinction in circuit breakers," *AIEE Transactions* **78**(Part III), 1508–1517 (1959).

Browne, Jr., T. E., "Practical modelling of the circuit breaker Arc as a short-line-fault interrupter," *IEEE T-PAS* **97**, 838–847 (1978).

Browne, Jr., T. E., ed., *Circuit Interruption*, Chapters 4, 6, Appendix Marcel Dekker, New York, 1984.

Cassie, A. M., Chapter 2, Introduction to the theory of circuit interruption, in H. Trencham ed., *Circuit Breaking*, Butterworth Scientific Publications, 1953.

Cassie, A. M., "Arc rupture and circuit severity: A new theory," Report No. 102, CIGRE, 1939.

CIGRE Study Committee 13, "Practical application of arc physics in circuit breakers: Survey of calculation methods and application guide," Report 118, Report of Working Group 01 *Electra*, 65–79, 1988.

CIGRE SC 13, "Application of black box modelling to circuit breakers", Report No. 149, Report of WG 011993, *Electra*, 41–71, 1993.

CIGRE SC 13, "Interruption of small inductive currents," Report No. 72, Chapters 1, 2, 13, Report of WG 01, *Electra*, 73–103, 1980.

CIGRE WG 13–01, "Analytical and graphical tools for circuit breaker behaviour description," CIGRE Report of WG 13-01, (SC) 27 IWD, 13–97, 1997.

Garrard, C. J. O, "High-voltage switchgear" *Proc. IEE* **123**, (10R), 1053–1080 (1976).

Garzon, R. D., *High-Voltage Circuit Breakers*, Chapter 1, 4, 5, Marcel Dekker, New York, 1997.

Heroin, P. *et al.*, "Indirect testing of an air-blast circuit breaker. Statistical comparison with direct testing," Report 144, CIGRE, 1966.

Kopplin, H, "Mathematische Modelle des Schaltlichtbogens", *ETZ Archiv* **H7**, S209–S213 (1980).

Lee, T. H., *Physics and Engineering of High Power Switching Devices*, Chapters 6, 9, MIT Press, Cambridge, USA, 1975.

Mayer, H., "High-voltage circuit breakers," *Electra* (Jubilee issue), 95–106 (1972).

Mayr, O., "Beitraege zur Theorie des statischen und dynamischen Lichtbogens", *Archiv für Elektrotechnik* **37**(H12), S588–S608 (1943); Mayr, O., "Über die Theorie des Lichtbogens und seine Loeschung", *ETZ-A* **64**, S645–S652 (1943).

Möller, K., "Die Geschichte der Mayrschen Differentialgleichung des Dynamischen Lichtbogens", *Tech. Mitt. AEG-Telefunken*, Jg. **62**(7), S338–S341 (1972).

Nakanishi, K., ed., *Switching Phenomena in High-Voltage Circuit Breakers*, Chapter 1, 2, 3, Marcel Dekker, New York, 1991.

Portela, C. M. *et al.*, Circuit breaker behaviour in reactor switching. applicability and limitations of the concept of chopping number," *IEEE T-PWRD* **3**(3), 1009–1021.

Ravindranath, B. and Chander, M., *Power System Protection and Switchgear*, Chapters 1, 13, 14, 16, Wiley Eastern, New Delhi, 1977.

Reece, M. P., in C. H., Flurscheim, ed., *Power Circuit Breakers: Theory and Design*, Chapters 1, 2, Institution of Electrical Engineers, 1982.

Rieder, W., "Arc-circuit-interaction near current zero and circuit-breaker testing," *IEEE International Symposium on High-Power Testing*, 705–713 (1971).

Rieder, W., "Circuit breakers: physical and engineering problems, Part I: Fundamentals," *IEEE Spectrum* (July), 35–43 (1970); Part II: Design considerations, *IEEE Spectrum* (August), 90–94 (1970); Part III: Arc-medium considerations, *IEEE Spectrum* (September), 80–84 (1970).

Rieder, W. und Pratl, J., "Das Löschen und Wiederzünden von Schaltlichtbögen," *Scientia Electrica* **11**(2), S33–S48 (1965).

Rieder, W., *Plasma und Lichtbogen*, Chapters 5, 6, Vieweg & Sohn GmbH, Braunschweig, Germany, 1967.

Rieder, W., "The activities of the current zero club," *Electra* **58**, 25–32 (1992).

Reid, W. E., "Effect of transient recovery voltage (trv) on power circuit interruption" *IEEE Power System Transient Recovery Voltages*, Chapter 1 course # 87TH0176-8-PWR (1987).

Rizk, F., "Arc instability and time constant in air-blast circuit breaker," Report 107 CIGRE, 1964.

Ruedenberg, R.,*Transient Performance of Electric Power Systems: Phenomena in Lumped Networks*, Chapter 39,, McGraw-Hill, New York, 1950.

Saha, M. N., "Ionisation in the solar chromosphere," *Philos. Mag.* **40**(472), (1920).

Schavemaker, P. H. and van der Sluis L., "The influence of the topology of test circuits on the interrupting performance of circuit breakers," *IEEE Transactions on Power Delivery* **18**(4), 1822–1828 (1995).

Schneider, J., "Beschaltungsmasznahmen zur Erhöhung des thermischen Ausschaltvermögens von Hochspannungsleistungsschaltern" *ETZ Archiv* **5**(H5), S149–S153.

Simon, H. Th., "Über die Dynamik der Lichtbogenvorgänge und über Lichtbogen-Hysteresis.", *Physikalische Zeitschrift* **6**(10), S297–S319 (1905).

Slamecka, E., *Pruefung von Hochspannungs-Leistungsschaltern*, Part IV, Chapter 6, Springer-Verlag, New York, 1966.

Slamecka, E. and Waterschek, W., Schaltvorgaenge in Hoch-und Niederspannungsnetzen, Chapter 3, Siemens Aktiengesellschaft, Berlin, 1972.

Slepian, J., "Extiction of an A-C. arc," *AIEE Trans.* **47**, 1398–1408 (1928).

CIGRE Study Committee 13, WG 01, "Application of black box modelling to circuit breakers," *Elektra* **149**, 41–71 (1993).

Sorensen, R. W. and Mendenhall, H. E. " Vacuum switching experiments at California institute of technology", *AIEE Trans.* **45**, 1102–1105.

van der Sluis, L., Rutgers, W. R., and Koreman, C. G. A., "A physical arc model for the simulation of current zero behaviour of high-voltage circuit breakers," *IEEE T-PWRD* 7(2), 1016–1022 (1992).

# 5

# Switching Transients

When load break switches, circuit breakers, disconnectors, or fuses operate, a switching action takes place in the network and parts of the power system are separated from or connected to each other. The switching action can be either a closing or an opening operation in the case of a switching device. Fuses can perform opening operations only. After a closing operation, transient currents will flow through the system, and after an opening operation, when a power-frequency current is interrupted, a *transient recovery voltage* or *TRV* will appear across the terminals of the interrupting device. The configuration of the network as seen from the terminals of the switching device determines amplitude, frequency, and shape of the current and voltage oscillations. When capacitor banks for voltage regulation are placed in a substation, the switching devices interrupt a mainly capacitive load when operating under normal load conditions. The current and voltage are approximately $90^\circ$ out of phase and the current is *leading* the voltage. When a large transformer is disconnected in a normal load situation, current and voltage are also approximately $90^\circ$ out of phase but now the current is *lagging*. Closing a switch or circuit breaker in a dominantly capacitive or inductive network results in *inrush currents*, which can cause problems for the protection system.

The *short-line fault* is of special importance for $SF_6$ circuit breakers. A fault is called a *short-line* fault when the short-circuit, usually a single line-to-ground fault, occurs on a high-voltage transmission line, a few hundred meters to a few kilometres from the breaker terminals. A very steep triangular-shaped oscillation immediately after current zero puts stress on the still-hot arc channel and can easily cause a *thermal breakdown*.

## 5.1 INTERRUPTING CAPACITIVE CURRENTS

Power systems contain lumped capacitors such as capacitor banks for voltage regulation or power factor improvement and capacitors that are part of filter banks to filter out higher harmonics. In addition, cable networks on the distribution level form a mainly capacitive load for the switching devices. Capacitive switching requires special attention because, after current interruption, the capacitive load contains an electrical charge and can cause a dielectric re-ignition of the switching device. When this process repeats, the interruption of capacitive currents causes high overvoltages. Figure 5.1 shows a single-phase representation of a capacitive circuit. The inductance $L_s$ is the inductance of the supply and represents the synchronous inductance of the supplying generators and the leakage inductance of the power transformers. $R_s$ and $C_s$ generate together the supply side TRV and represent the damping and the capacitance of the voltage transformers, current transformers, bus bars, and so forth. The capacitive load is represented by a lumped capacitor $C$, connected via a stray inductance $L'$ with the load side of the circuit breaker. When $L_s \gg L'$ and $\omega L_s \ll 1/\omega C$, the current is *leading* the supply voltage by 90°.

The capacitive current is small, a few Amperes to several hundreds of Amperes, compared with the rated short-circuit current for which the circuit breaker is designed, and the capacitive current can be interrupted even at small arcing times. At the instant of current interruption, the capacitor is fully charged and the voltage is approximately equal to the peak voltage of the supply. After half a cycle, the supply voltage has reversed its polarity and the voltage across the breaker terminals is twice the peak value of the supply voltage, as can be seen in Figure 5.2.

When the circuit breaker is in a closed position, the voltage behind the breaker is higher than the supply voltage. The voltage difference is $\Delta U = U_c - E$. This is called the *Ferranti rise* and can also be seen as the effect of the capacitor acting as a source of reactive power; it subsequently

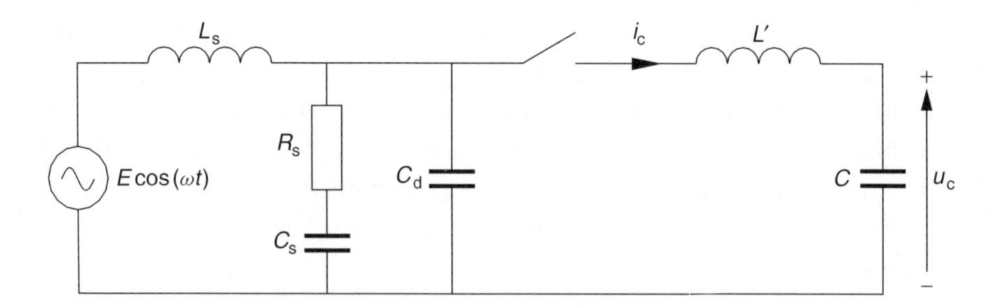


**Figure 5.1** Single-line representation of a capacitive circuit

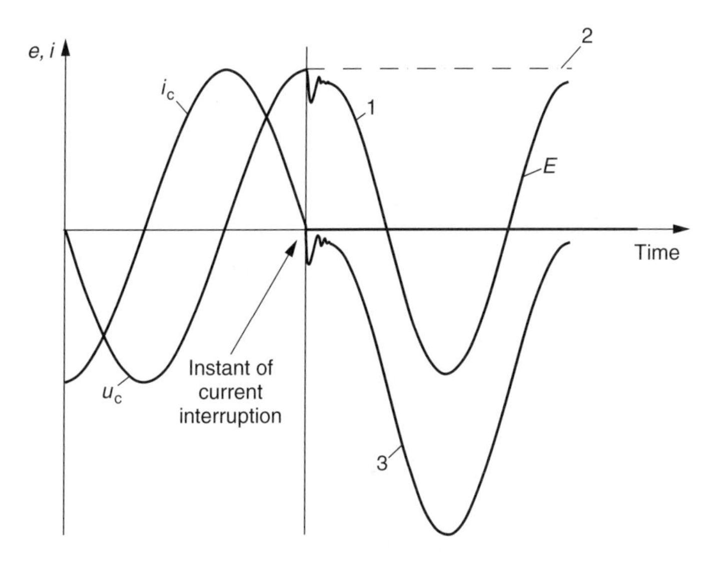


**Figure 5.2** Current and voltage traces during the interruption of a capacitive current 1: supply voltage, 2: voltage on the capacitor, and 3: voltage across the circuit breaker

increases the voltage level. As a consequence of the Ferranti rise, a voltage jump occurs in the voltage at the supply side of the breaker. The frequency of the transient oscillation as a result of this voltage jump is

$$f = \frac{1}{2\pi\sqrt{L_s C_s}} \tag{5.1}$$

and the rate of rise of this voltage jump can be calculated with

$$\frac{\mathrm{d}u}{\mathrm{d}t} = Z_s \frac{\mathrm{d}i_C}{\mathrm{d}t} \tag{5.2}$$

in which

$$Z_s = \sqrt{\frac{L_s}{C_s}} \tag{5.3}$$

and can be regarded as the characteristic impedance of the supply circuit.

In the case that $C$ is a 10 MVAR capacitor bank for power factor compensation in a 20-kV substation with a 50-Hz short-circuit level of 1100 MVA, the short-circuit current is 31.5-kA rms.

The capacitive current is

$$i_c = (10\ \text{MVAR}/20\ \text{kV})/\sqrt{3} = 290\ \text{A} \tag{5.4}$$

The impedance of the capacitor bank is

$$X_c = (20\ \text{kV}/\sqrt{3})/290 = 40\ \Omega \tag{5.5}$$

and the value of the capacitance is 80 μF. The short-circuit power gives us the value of the supply inductance:

$$L_s = \{(20\ \text{kV}/\sqrt{3})/31.5\ \text{kA}\}/100\,\pi = 1.2\ \text{mH} \tag{5.6}$$

When the supply side capacitance $C_s$ is 100 nF, the Ferranti rise is 9.2 percent and the frequency of the transient voltage jump is

$$f = \frac{1}{2\pi\sqrt{L_s C_s}} = 14.7\ \text{kHz} \tag{5.7}$$

In systems where distribution transformers are used with a rather high leakage inductance to reduce the short-circuit power, the supply inductance $L_s$ has a rather high value and the voltage jump due to the Ferranti rise is relatively big. In minimum-oil circuit breakers, the steep rate of rise of the transient voltage jump can cause a thermal re-ignition of the recovering arc channel and thus extends the arcing time of the breaker with another half cycle. The small capacitive current continues to flow and causes a high arc voltage in the oil. Because of this, there is considerable energy input in the arc column, which can make the minimum-oil circuit breaker to explode.

The voltage transient is a source-side phenomenon. At the load side, where the capacitor bank or the cables are connected, there is no change in voltage. The trapped charge causes a constant voltage, but when the supply voltage has the opposite polarity, nearly two times the peak value of the supply voltage is present across the breaker contacts. When we realise that the small capacitive current could already be interrupted at a short arcing time and that the arcing contacts therefore have not travelled very far, the gap between the arcing contacts is still narrow and a dielectric breakdown of the extinguishing medium can occur. When the re-ignition occurs (see Figure 5.3), the capacitance $C$ discharges itself via the re-ignited arc channel and the inductances $L_s$ and $L'$. The result is an oscillating current with the following frequency and peak value.

$$f = \frac{1}{2\pi\sqrt{L_s C}} \tag{5.8}$$

for

$$L_s \gg L'$$

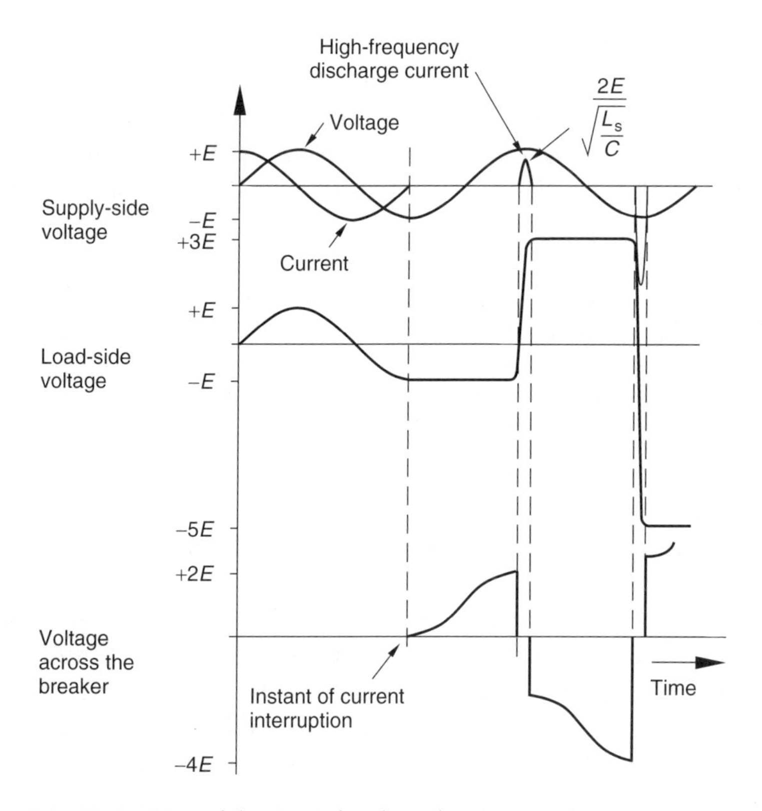


**Figure 5.3** Re-ignition of the circuit breaker when interrupting a capacitive current

and a peak value

$$I = \frac{2E}{\sqrt{\frac{L_s}{C}}} \tag{5.9}$$

The smaller the value of $L_s$, the higher the frequency and larger the amplitude of the transient current. When at the instant of re-ignition the value of the voltage on the capacitor $C$ was $-E$, the voltage is then $+3E$ at the first zero crossing of the transient current. When the arc channel extinguishes, the circuit breaker interrupts the oscillating current. The recovery voltage across the contact gap has now increased to $4E$ (the contribution of the supply-side voltage is $-E$ and of the trapped charge on the capacitor $C$ is $+3E$). The arcing contacts of the circuit breaker have parted only a little further and another re-ignition is likely to occur. The result of this second re-ignition of the extinguishing medium is an oscillating current with a doubled amplitude, and when the breaker interrupts the transient current at its first zero crossing again, the voltage on capacitor $C$ has increased to $5E$ and the voltage across the breaker contacts is now $6E$. When a couple of re-ignitions occur in this way, very high voltages build up across the interrupting chamber, and it is most likely that a flashover on the outside of the interrupter chamber takes place. The circuit breaker is short-circuited out of the system in this way and cannot function any more; this is a very dangerous and an unwanted situation. A dielectric breakdown of the arc channel within 5 ms after interruption of the capacitive current is called a *re-ignition* and a dielectric breakdown of the arc channel after 5 ms is called a *restrike*. High-voltage circuit breakers, which have to perform capacitive current switching, should be *restrike-free* to avoid overvoltages.

When a circuit breaker switches on a capacitive load (performing a closing operation in the circuit of Figure 5.1), it is possible that a breakdown of the extinguishing medium occurs when the supply voltage is at its maximum just before the contacts touch and when, for instance, the capacitive load contains a trapped charge with a polarity opposite to the polarity of the supply. This breakdown of the extinguishing medium is called a *prestrike*, and the suddenly created plasma channel causes a shock wave in the extinguishing medium. This is a dangerous situation especially for minimum-oil circuit breakers, because of their small oil volume and the incompressibility of oil, when the shock wave can destroy the arcing chamber. For air-blast and $SF_6$ circuit breakers, this is less likely to occur because gas is more compressible than a liquid.

A very specific capacitive switching duty is *back-to-back switching*. In this situation, as depicted in Figure 5.4, a large capacitor bank is not only connected at the load side of the breaker but also at the supply side.

When switching the capacitive current in a back-to-back situation, in the most onerous situation, a re-ignition occurs when the load-side capacitor has a polarity opposite to that from the supply-side capacitor. The discharge current then flows from one capacitor bank to the other via the stray inductance (caused by the loop of the bus bars and the connecting wires). The frequency of the current oscillation and the peak of

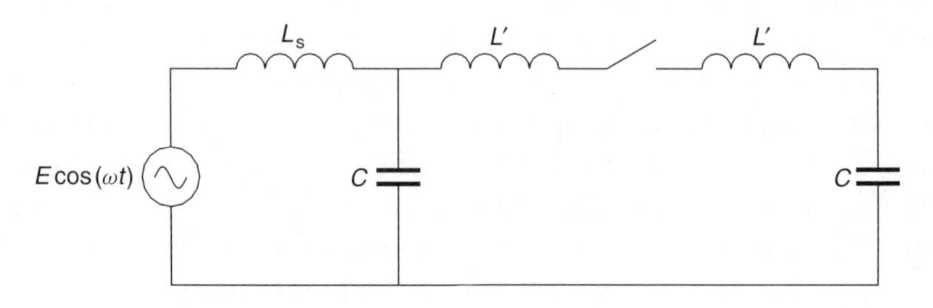


**Figure 5.4** Single-line representation of a back-to-back situation

the current are therefore much higher than in the case of single-capacitor bank switching.

Switching *unloaded high-voltage transmission lines* in and out of service is, in principle, the same as switching a capacitor bank because an unloaded transmission line has a dominantly capacitive behaviour. In addition, a small voltage jump occurs because of the Ferranti-rise effect. High-voltage transmission lines also have *travel times* and must therefore be represented by *distributed elements* instead of *lumped elements*. When a transmission line is energised after closing of a switching device, the resulting voltage wave reflects and causes doubling of the voltage at the open end of the line. In addition, switching off an unloaded transmission line is, in principle, similar to switching off a capacitor bank. The Ferranti rise causes a transient voltage jump that results in a voltage wave travelling along the line. The interrupting device is stressed by a voltage jump at the supply side and an oscillation at the line side. The resulting steep voltage jump across the contacts of the interrupting device can cause, because of the short arcing time, a re-ignition and prolong the arcing time.

When we analyse three-phase switching of unloaded high-voltage transmission lines, the capacitive coupling between the phases and the capacitance to ground have to be taken into account (see Figure 5.5).

When the first phase (e.g. phase 1) has cleared the power-system, voltage $V_i$ of the neighbouring phases 2 and 3 is coupled into the line-side DC voltage $V_1$ with a coupling factor $k = V_i/V_1$. The coupling factor between the neighbouring phases and phase 1 depends on the ratio of $C_{21}/C_{1\ \text{earth}}$ and $C_{31}/C_{1\ \text{earth}}$. When the second phase clears (let us assume that this is

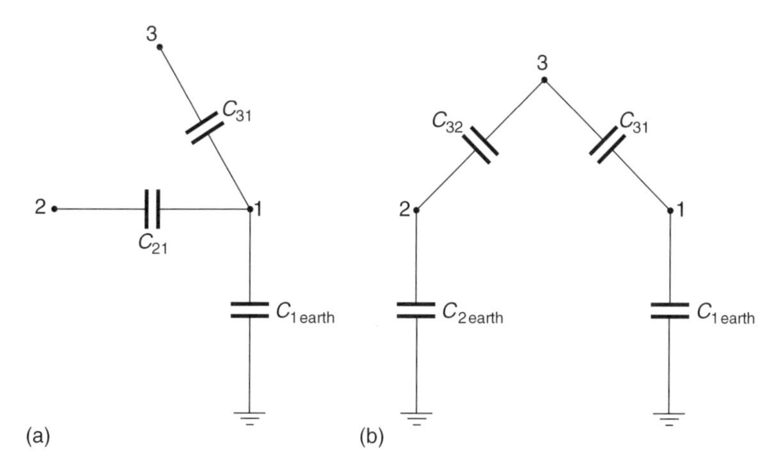


**Figure 5.5** The coupling effect of the line capacitances when switching unloaded high-voltage transmission lines (**a**) phase 1 has been switched off and phases 2 and 3 are still at voltage (**b**) phases 1 and 2 are switched off and phase 3 is still at voltage

phase 2), then phase 3 is still at system voltage and this voltage couples into the line-side DC voltages of phase 1 and phase 2. The value of the coupling factor depends on the tower structure and circuit design but usually has a value between 0.2 and 0.4. For a double circuit, when two circuits are hanging in the same tower, the coupling factor can be higher when the neighbouring circuit is in operation. Capacitive coupling increases the TRV across the contacts of the interrupting device with the coupling factor (Figure 5.6).

When unloaded high-voltage transmission lines are switched off, a combination of the voltage jump at the supply side, the transient voltage oscillation at the supply side, the voltage oscillation at the line side, and the capacitive coupling with the neighbouring phases can result in a recovery voltage for the first phase to clear, which can be as high as 2.8 per unit (Figure 5.7).

When switching *unloaded high-voltage cables* only the capacitive load current is interrupted. This current is small compared with the capacitive current when switching capacitor banks, but large compared with the capacitive current when switching unloaded high-voltage transmission lines. The actual construction of the cable is of importance when we want to study the transient recovery voltage in detail. For three-phase cables with the three conductors in a lead sheet with an earth screen, the same coupling effects as with transmission lines occur.

When each conductor has its own earth screen, only the capacitance to earth plays a role and the resulting TRV has the same shape as the

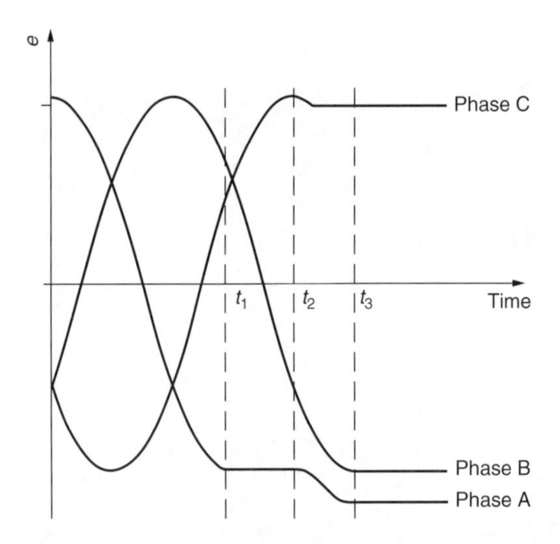


**Figure 5.6** Line-side voltages after switching off an unloaded high-voltage transmission line. The three phases interrupt the capacitive current at $t_1$, $t_2$, and $t_3$

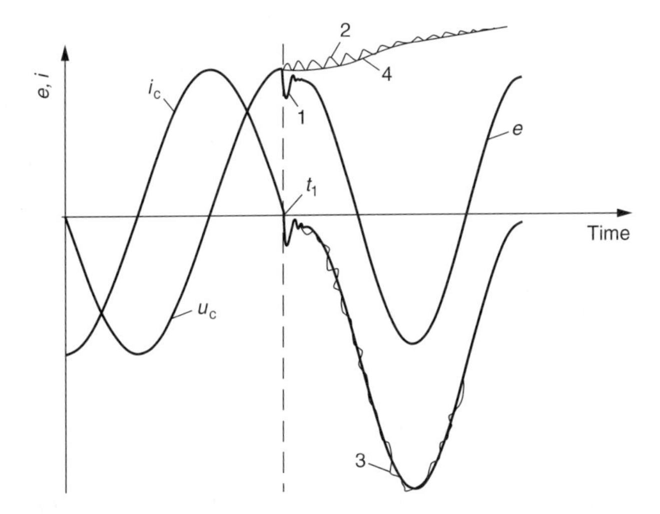


**Figure 5.7** Current and voltage traces after switching off an unloaded high-voltage transmission line: 1: voltage jump at the supply side; 2: transient voltage oscillation at the supply side; 3: voltage across the contacts of the interrupting device; 4: power frequency voltage coupled in from the neighbouring phase

TRVs in the case of switching capacitor banks with an earthed neutral in a grounded system.

## 5.2 CAPACITIVE INRUSH CURRENTS

In a power system, capacitor banks for load factor improvement or for filtering out higher harmonics have to be switched in and out of service regularly. The interruption of a capacitive current can cause dielectric problems for the switching device, but when a capacitor bank is taken into service, large inrush currents can flow through the substation. Figure 5.8 shows the single-line diagram of a substation with a capacitor bank for

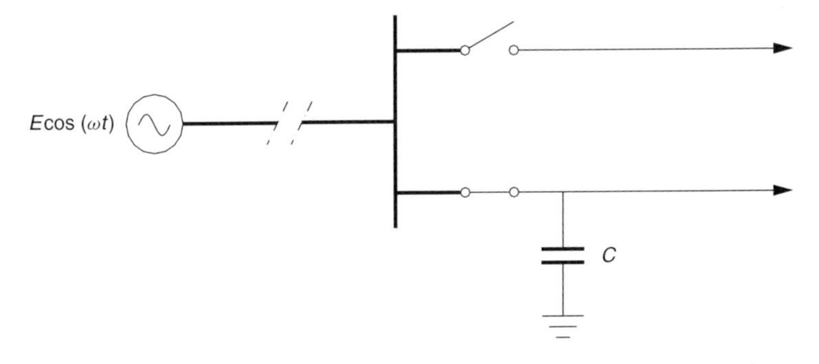


**Figure 5.8** Single-line diagram for a substation with a capacitor bank for voltage regulation

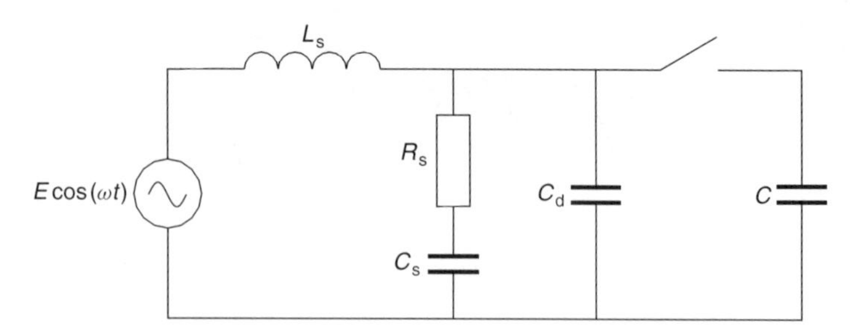


**Figure 5.9** Single-phase lumped-element representation of a substation with a capacitor bank for voltage regulation

voltage regulation. The single-phase representation with lumped elements is depicted in Figure 5.9.

In Chapter 1, *Basic Concepts and Simple Switching Transients*, the transient current that is flowing after a source has been switched on in an LC series network has been analysed. The transient current $i(t)$ depends on the initial conditions in the circuit and the circuit parameters. If capacitor C is uncharged at $t = 0$, when the switching device closes, the expression for the current is

$$i(t) = \left[\frac{E(0)}{Z_s}\right] \sin(2\pi f_s t) \tag{5.10}$$

where $Z_s = \sqrt{(L_s/C)}$ is the characteristic impedance of the circuit, $f_s = 1/2\pi\sqrt{(L_s/C)}$ is the natural frequency of the circuit, and $E(0)$ is the instantaneous voltage of the supplying source at the instant of closing of the switching device at $t = 0$.

It is assumed that the power frequency remains at a constant value during the time period when the transient currents are flowing, as the frequency of the transient is considerably higher than the power frequency. If for reactive power compensation, to maintain the voltage level within its limits, a capacitor bank of 10 MVAR is connected at the 20-kV bus of a distribution substation at 50 Hz with short-circuit rating of 31.5 kA rms, the rms value of the capacitive current in the steady situation is (see Equation (5.4)) $i_{cap} = 290$ A, the impedance of the capacitor is (see Equation (5.5)) $X_C = 40\ \Omega$ corresponding to a capacitor value $C = 80\ \mu$F. From the voltage level and the short-circuit inductance of the source (see Equation (5.6)), $L_s = 1.2$ mH can be calculated. The characteristic impedance of the circuit of Figure 5.9 is therefore $Z_s = 3.8\ \Omega$ and the natural frequency $f_s = 514$ Hz.

When at the instant of closing the switch at $t = 0$ the source voltage is at its maximum, the peak of the transient current (Equation (5.10)) is 4300 A. So the peak of the transient inrush current is more than ten times the value of the peak of the steady-state capacitive current. These rather high inrush currents can cause damage to the capacitors of the capacitor bank. But the switching device can also be damaged. When a dielectric breakdown occurs just before the contacts touch during the closing operation (a so-called *prestrike*), the rather large inrush current can cause the contact material to meld. Especially in cases in which vacuum circuit breakers are used to switch the capacitor banks, the contacts might weld together. A prestrike in a minimum-oil circuit breaker or in an $SF_6$ circuit breaker can do other damages to the breaker as has already been mentioned in Section **5.1.** The suddenly created plasma channel causes a shock wave and can make, in the case of a minimum-oil circuit breaker, the complete interrupter to explode or, when $SF_6$ is the extinguishing medium, can sometimes damage the arcing chamber or nozzle.

## 5.3 *INTERRUPTING SMALL INDUCTIVE CURRENTS*

When an interrupting device interrupts a *small inductive current*, the current can, like a small capacitive current, be interrupted at a short arcing time. Interrupting devices such as high-voltage circuit breakers are designed to clear a large short-circuit current in milliseconds so that the cooling of the arc maintained by a small current is easy. The gap between the arcing contacts, after current interruption, is rather small and the capability to withstand dielectric breakdown is relatively low. Small inductive currents occur when unloaded transformers are taken in and out of service, motors are disconnected, or electric furnaces are switched. A single-line representation of a circuit for small inductive currents is given in Figure 5.10.

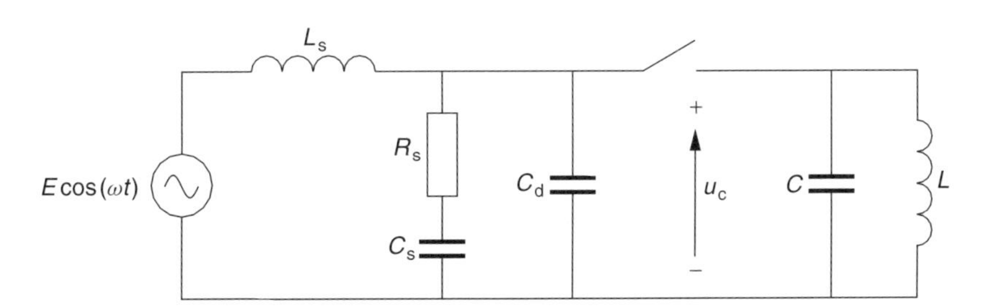


**Figure 5.10** Single-phase lumped element representation of a circuit for small inductive currents

The inductivity $L$ of the load is dominant, which means that the load makes the current to lag the voltage. The capacitance $C$ of the load is usually very small: a few nanoFarads for a distribution transformer and a few picoFarads for an air-core reactor, depending on the design. For a motor or an electric oven, connected to the supply by means of a cable, the value of the capacitance $C$ is determined by the length and type of the cable.

When the small inductive current is interrupted, commonly at a short arcing time, the load capacitance $C$ is charged at

$$u_c = L\,di/dt \tag{5.11}$$

which is a value close to the supply voltage, because $L \gg L_s$. After current interruption, $C$ discharges itself through $L$ by means of an oscillating current with frequency

$$f_1 = \frac{1}{2\pi\sqrt{LC}} \tag{5.12}$$

The frequency of this oscillation can be from a few kiloHertz to several kiloHertz and thus creates very steep $du/dt$'s at the load side of the switching device. It is likely that the steep increase of the TRV at the load will cause a dielectric breakdown of the extinguishing medium between the narrow contact gap, and a re-ignition occurs. After a re-ignition, the inductive load is connected with the supply again and the capacitance $C$ of the load is charged by a current with frequency

$$f_2 = \frac{1}{2\pi\sqrt{L_s C}} \tag{5.13}$$

and a high-frequency current flows through the arc channel. When the switching device interrupts this high-frequency current, capacitance $C$ is charged to a higher value because the $di/dt$ of the re-ignition current is considerably higher than the power frequency current, which was interrupted first (Equation (5.11)). The capacitance $C$ will discharge itself again through the inductivity $L$. In the mean time, the arcing contacts have moved further apart and have increased the capability to withstand dielectric breakdown of the contact gap. However, the voltage stress caused by the oscillation at the load side has increased as well and another re-ignition might occur. When this process repeats itself a few times, a considerable overvoltage can be the result.

When switching small capacitive currents or small inductive currents, the unwanted result is in both cases the same: high overvoltages (and possible damage to high-voltage equipment in the vicinity) or even a flashover on the outside of the switching device. In the latter case, the high-voltage circuit breaker or the load breaker switch is short-circuited out of the system – a dangerous situation. Apart from the current being leading or lagging, the principle difference between the two switching cases is that the capacitance $C$ at the load side, in the case of small inductive current switching, does not remain charged but discharges itself through the load $L$.

## 5.4 TRANSFORMER INRUSH CURRENTS

Power transformers bring in the transmission system the electrical energy to higher voltage levels to reduce losses when transporting electrical energy over long distances, and at the distribution level they transform the voltage down to the required level for the consumer. A switching operation carried out in a substation always involves transformer switching. When switching off a transformer under load, the load determines the power factor and the switching device interrupts the load current, which is normally not a problem for the switch or circuit breaker, creating no overvoltage in the system. When a part of the system is energised by connecting it to the rest of the system by a closing operation of a switch or breaker, the transformer can cause high inrush currents. The nonlinear behaviour of the *transformer core* is the cause of this. An air-core reactor switched on to compensate for cable charging currents does not cause inrush currents.

When a power transformer is energised under no-load condition, the *magnetising current* necessary to maintain the magnetic flux in the core is in general only a few percent of the nominal rated load current. Figure 5.11 shows the *magnetising curve* and the *hysteresis loop*.

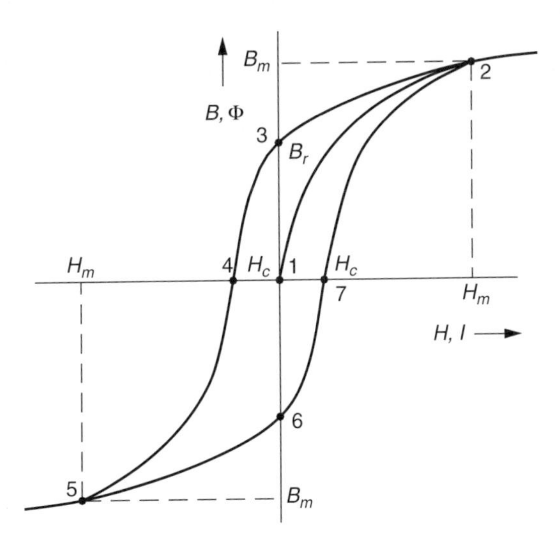


**Figure 5.11** Magnetising curve and hysteresis loop of a transformer core

Starting with an unmagnetised transformer core the flux density $B$ follows the initial magnetisation curve, starting at the origin (at 1), as the magnetic field intensity $H$ is increased to a value $H_m$ where the curve flattens off and saturation is reached (at 2). When $H$ is reduced to zero, $B$ does not go to zero but has a *residual flux density* or *remanence* $B_r$ (at 3). When $H$ reverses because the polarity of the current changes, $H$ increases negatively and $B$ comes to zero at a negative field called the *coercive force* $H_c$ (at 4). As $H$ increases still more in the negative direction, the transformer core becomes magnetised further with negative polarity. The magnetisation at first is easy and then gets harder as saturation is being reached. The magnetic field intensity equals $-H_m$ (at 5). When the applied field goes to zero again, the core is left with a residual magnetisation with flux density $-B_r$ (at 6). When $H$ reverses at the next current loop and increases in the positive direction, $B$ comes to zero at a positive field or coercive force $H_c$ (at 7). With further increase, the transformer core reaches saturation with the original polarity.

When a power transformer has been switched off from the system, the transformer core is left with residual flux $B_r$. When the power transformer is connected to the network again at such an instant that the polarity of the system voltage is the same as the polarity of the residual flux $B_r$, then at maximum voltage, the total flux density in the core would have increased

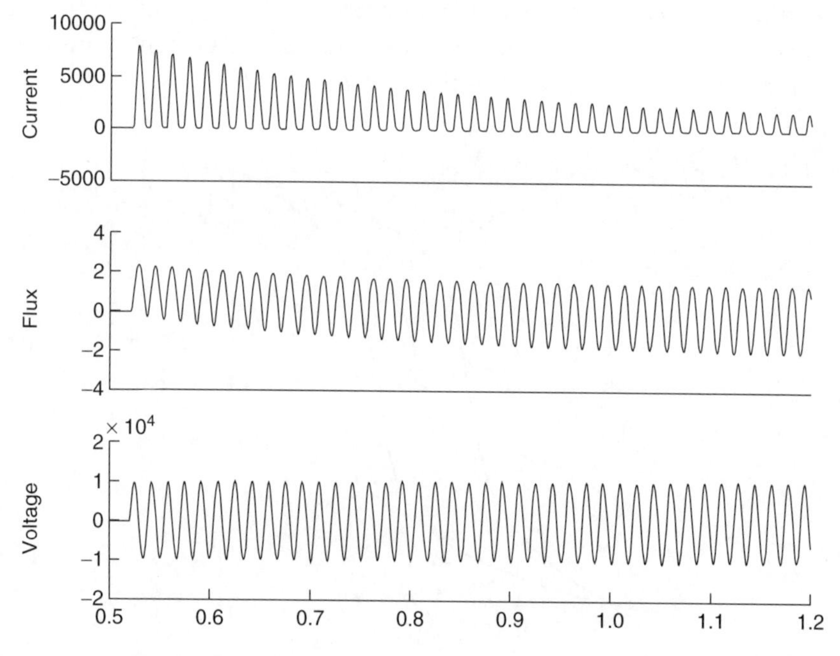


**Figure 5.12** Transformer inrush current, the flux in the core, and the supply voltage

to $B_m + B_r$. The core is forced into saturation and the transformer draws a large current from the supplying network. When the voltage reverses its polarity in the next half cycle, then the maximum flux in the core is less than the maximum flux density $B_m$ in the no-load situation. The transformer inrush current is therefore asymmetrical and also contains a DC component, which takes seconds to disappear (Figure 5.12).

## *5.5 THE SHORT-LINE FAULT*

In the 1950s, it was observed that minimum-oil circuit breakers, in particular, exploded after clearing short-circuit currents with a current level below the nominal rated short-circuit breaking capability. In addition, air-blast breakers occasionally failed to interrupt short-circuit currents considerably smaller than the current values they were designed and tested for. Careful analysis has revealed that in many cases the fault occurred on a high-voltage transmission line, a few kilometres from the circuit breaker terminals. The *short-line fault* was discovered, and several years later the short-line fault test appeared in the standards. Presently, the short-line fault test is considered to be one of the most severe short-circuit duties for high-voltage circuit breakers. The short-line fault test creates, after current interruption, a very steep voltage oscillation at the line side of the breaker. This puts a high stress on the still-hot arc channel, and this can cause a *thermal breakdown* of the arc channel. The arcing time is prolonged, and when the same happens at the next current zero, the breaker fails to interrupt the short-circuit current. $SF_6$ circuit breakers are especially very sensitive for short-line faults occurring only a few hundred metres to a few kilometres from the breaker terminals. Air-blast breakers are, in particular, stressed by faults occurring further away, several kilometres from the breaker terminals. The reason for this will be explained later in this paragraph; first the TRVs at the supply side and later the TRVs at the line side of the breaker will be analysed. Figure 5.13 shows the single-phase representation of a short-line fault.

After interruption of the short-line fault current, the TRV builds up across the breaker contacts. This TRV consists of a voltage generated by the supply network and of a voltage created by the line-side oscillation. The TRV from the supply network has a more or less (1-cosine) shape, which we can expect because it is basically an oscillating LCR-series network. The line-side oscillation has a triangular-shaped or saw-tooth-shaped waveform. The TRVs are depicted in Figure 5.14.

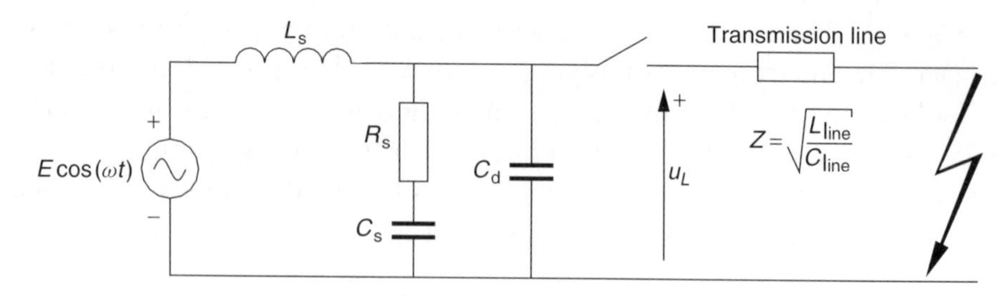


Figure 5.13 The short-line fault

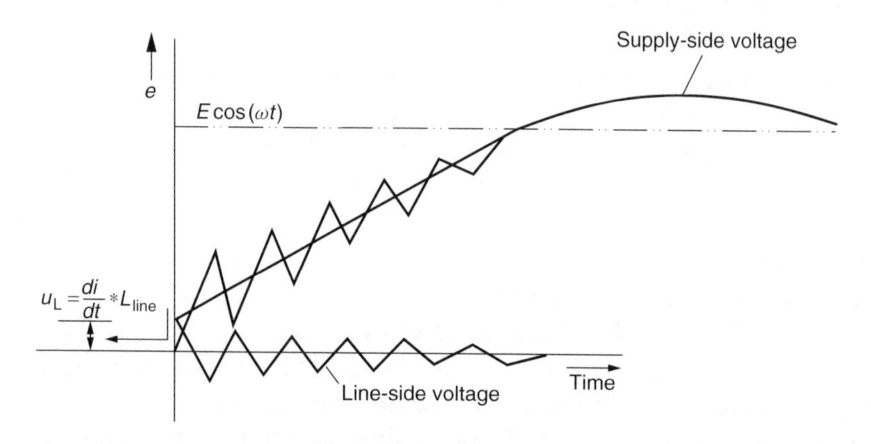


Figure 5.14 Line-side, supply-side TRVs and the TRV across the breaker contacts

The circuit breaker interrupts the short-circuit current at current zero. The rms value of the current is less than the current in the case of a bolted terminal fault directly behind the terminals of the breaker. The impedance of the short-circuited length of line reduces the current. Because of the inductive nature of the circuit, the supply voltage $E\cos(\omega t)$ is close to its peak value, and the voltage at the line side has the value $u_L = L_{line}di/dt$. The voltage profile over the line decreases linearly from $u_L$ to zero, when we consider a single line-to-ground fault, and the line is therefore left charged at the instant of current interruption. The charge is maximal at the side of the breaker and zero at the fault place. The charge on the line will balance itself and this balancing of charges causes travelling waves on this part of the high-voltage transmission line. The reflections of the travelling waves against the short-circuit point and the open end at the breaker side cause a triangular-shaped waveform. (See also Chapter 3, *Travelling waves,* Section 3.5.)

For a single line-to-ground fault at a distance $l$ from the breaker, the travel time for the electromagnetic waves from the breaker to the fault place is $\tau_{\text{line}} = 1/\nu$, where $\nu$ is the wave velocity fixed by the transmission line parameters. After twice the travel time, the wave reflected by the short-circuit point arrives at the open-line end near the breaker. This continuing process of reflection results in the triangular-shaped line-side oscillation. The first excursion of this line-side oscillation has a rate of rise, depending on the interrupted current level and the distance of the short-circuit point on the line, between 3 kV/μs and as much as 10 kV/μs. This very steep rate of rise of the TRV at the line side of the breaker has its impact on the still very hot gas mixture between the breaker contacts, after the arc plasma has vanished and the current has been interrupted. In a hot gas, there are always free electrons present as a result of the temperature and the extensive radiation in the arcing interval. The rapidly increasing electric field strength makes the free electrons drift, a so-called post-arc current flows between the breaker contacts in the still hot arc channel. When the cooling power of the interrupting process is such that the cooling prevails over the energy input as a result of the post-arc current and the rapidly increasing TRV, then a successful current interruption takes place. When the energy input in the hot gas channel prevails, a thermal re-ignition is the result and the circuit breaker has failed to interrupt the short-circuit current.

For circuit breakers there exists a so-called *critical line length*. The critical line length depends on the type of extinguishing medium. The critical line length, for instance, for $SF_6$ circuit breakers is around 93 percent and for air-blast breakers between 75 and 85 percent (see Figure 5.15). When the standards speak from a certain percentage of line fault, this means the reduction of nominal rated short-circuit current in the case of a bolted terminal fault by the impedance of the line segment at the load side of the breaker. When for example the 100 percent short-circuit rating of a circuit breaker is 40 kA and the short-circuited line reduces the short-circuit current till 36 kA, it is called a 90 percent short-line fault. The percentage of short-line fault can therefore easily be calculated with

$$\text{SLF percentage} = X_{\text{supply}}/(X_{\text{supply}} + X_{\text{line}}) \qquad (5.14)$$

The phenomenon of the *critical line length* can be understood by the following reasoning:

- When the line length increases (and the short-line fault percentage decreases), the current to be interrupted gets smaller.
- The impedance of the line increases.

- Because the rms value of the current gets smaller, the d$i$/d$t$ becomes lower
- The characteristic impedance $Z_{line}$ depends only on the line parameters and remains constant.
- The d$i$/d$t$ and also the d$u$/d$t$ of the line-side oscillation decreases because $\mathrm{d}u/\mathrm{d}t = Z_{line}\,\mathrm{d}i/\mathrm{d}t$.
- The increase in line length brings also an increase in line impedance, and therefore the amplitude of the triangular-shaped line-side oscillation, in spite of the lowering d$u$/d$t$, oscillates to a higher peak value.
- We now have two effects working opposite, lower rate of rise of the recovery voltage but a higher peak value. The difference in line length gives therefore a different stress on the arc channel after current interruption.
- Because the two mechanisms work opposite there is a maximum stress at a particular line length.

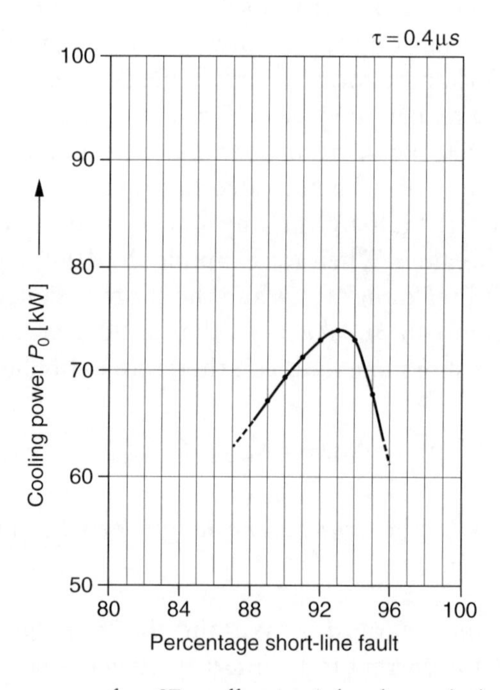


**Figure 5.15** Cooling power of an $SF_6$ puffer circuit breaker, calculated with a Mayr-arc model, required for a successful interruption at various short-line fault percentages

In all practical cases, the fundamental frequency of the line-side travelling waves is high compared with the frequency of the supply-side TRV, and therefore the voltage on the supply side changes only insignificantly during the time of the first excursion of the line-side oscillation. The amplitude of the first peak of the TRV can therefore be calculated without taking into account the supply-side oscillation.

When for a *three-phase short-line fault* the rate of rise of the line-side oscillation and the amplitude of the first peak are to be calculated, the supply network and the high-voltage transmission line are assumed to be symmetrical and both the neutrals of the supply and the fault are assumed to be solidly grounded. The influence of the circuit breaker arc and also the arc-circuit interaction are neglected. When calculations with travelling waves in three-phase systems have to be made, it is often suitable to apply the method of symmetrical components and divide the waves into phase components, different for the three phases, and into earth components, between neutral and ground, which are equal as Figure 5.16 shows.

The two types of components are related by

$$\begin{aligned} u_a + u_b + u_c &= 0 \\ u_e &= (U_a + U_b + U_c)/3 \end{aligned} \tag{5.15}$$

where $U_a$, $U_b$, and $U_c$ are the line-to-ground voltages, $u_a$, $u_b$, and $u_c$ the phase components and $u_e$ is the earth component. The line-to-ground

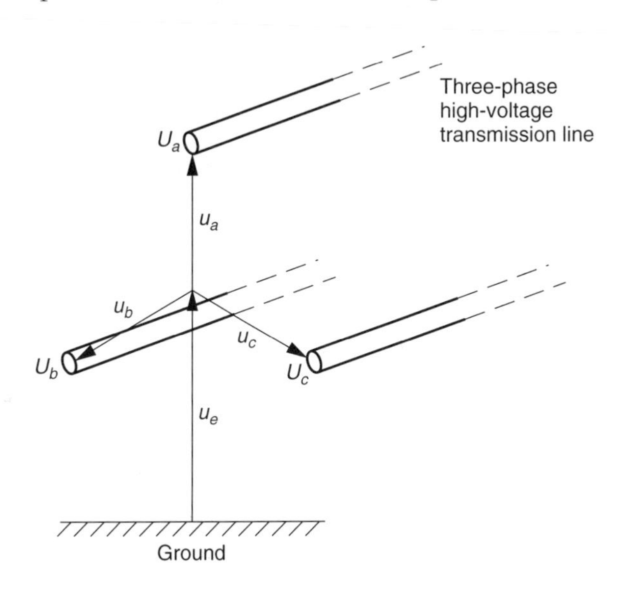


**Figure 5.16** Line-to-ground voltages $U_a$, $U_b$, and $U_c$ divided into phase components $u_a, u_b, u_c$ and earth component $U_e$

voltages can be written as

$$
\begin{aligned}
U_a &= u_a + u_e \\
U_b &= u_b + u_e \\
U_c &= u_c + u_e
\end{aligned}
\tag{5.16}
$$

The characteristic impedances for the travelling waves are

$$Z_a = u_a/i_a, \quad Z_b = u_b/i_b, \quad Z_c = u_c/i_c, \quad \text{and} \quad Z_e = u_e/i_e$$

The connection of the characteristic impedances, in general, in the case of the first phase to clear for a three-phase fault and the last phase to clear for a three-phase fault is shown in Figure 5.17.

For a symmetrical power system, where $Z_a = Z_b = Z_c = Z$, we know from the theory of the symmetrical components (see Chapter 2, *The Transient Analysis of Three-phase Power Systems*) that

$Z = Z_1 =$ the positive-sequence characteristic impedance (= the negative-sequence impedance)

$Z_e = (Z_0 - Z_1)/3$ with $Z_0$ being the zero-sequence characteristic impedance

The phase components of the line-to-ground voltages do not cause any earth current according to Equation (5.15) and are therefore not influenced

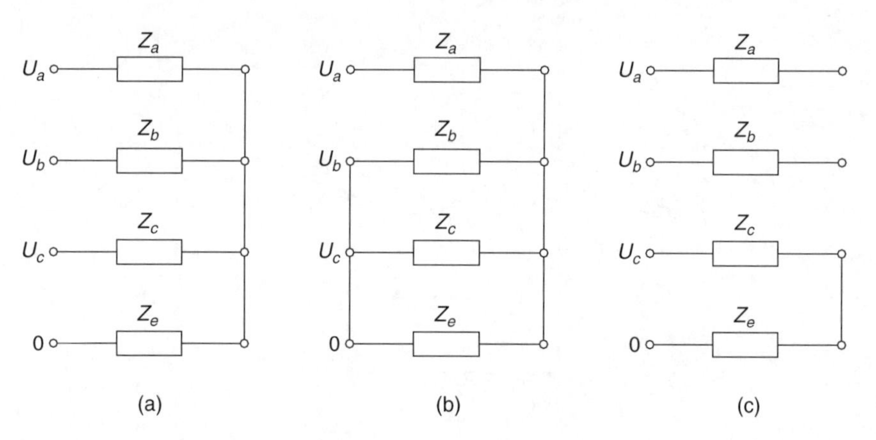


**Figure 5.17** Phase and earth components of the characteristic impedances. (a) general; (b) three-phase fault, first phase to clear; (c) three-phase fault, last phase to clear

by the conditions of the earth. On the other hand, the earth component of the line-to-ground voltages does give rise to currents that return through earth. The influence of the earth is, by the division in phase and earth components, clearly visualised, and we can now distinguish between earth waves and phase waves.

The phase waves are only attenuated by the resistance of the conductor. The voltage level of the travelling waves is usually below the level where corona occurs, and attenuation by corona losses does not take place. For line lengths of a few kilometres, the attenuation is rather small and the travelling waves can be regarded as undistorted waves, travelling with a velocity close to the speed of light. The earth waves, however, will be attenuated by the earth return path and this distorts the wave front. The distortion appears as a prolongation of the wave front in the order of one microsecond per kilometre and can therefore not be neglected.

For the first phase to clear and the last phases to clear, the conditions are different. For the *first phase to clear*, the characteristic impedances are connected as depicted in Figure 5.17. From Equation (5.15) it follows that

$$u_a + u_b + u_c = 0 \quad \text{and} \quad u_\mathrm{e} = 0 \tag{5.17}$$

Only phase waves are present and for a purely inductive line impedance the initial amplitudes will have, for the 50- or 60-Hz power frequency, the values

$$u_a = U_a = U_\mathrm{phase},\ u_b = U_b = -0.5\ U_\mathrm{phase},\ \text{and}\ u_c = U_c = -0.5\ U_\mathrm{phase} \tag{5.18}$$

The phase waves travel along the line, reflect against the short-circuit point, and travel backwards with the opposite polarity, and, in the case in which losses are neglected, they travel with the same amplitude. The conditions of Equation (5.17) are of course unchanged after the reflection, and the voltage at the breaker side of the line will continue to change with the same rate of rise until the reflected wave is back again. Then, when the reflected wave has reached the breaker terminal, phase waves and earth waves will be 'redistributed' because of the asymmetry (phase a open and the other two phases still arcing). This complicates the analysis, but as it plays a role after the appearance of the first peak of the line-side oscillation, it is of less importance in its influence on the thermal interruption interval of the circuit breaker. The rate of rise of the recovery voltage for the first clearing phase can be calculated from Figure 5.17

$$\frac{\mathrm{d}u}{\mathrm{d}t} = \frac{\mathrm{d}i}{\mathrm{d}t}\left(\frac{3Z_1Z_0}{Z_1 + 2Z_0}\right) \tag{5.19}$$

The rate of change of the current $\mathrm{d}i/\mathrm{d}t$ is the value just before current interruption and can be simply calculated (see Section 4.7 *Arc-circuit interaction*):

$$\mathrm{d}i/\mathrm{d}t = 2\pi f I\sqrt{2} \tag{5.20}$$

The time to the first peak of the line-side oscillation is twice the length of the short-circuited part of the transmission line divided by the wave velocity. The amplitude of the line-side TRV can be calculated from the $\mathrm{d}u/\mathrm{d}t$ and the time to peak. For a very accurate calculation of the amplitude, a correction should be made for the fact that for high-frequency currents the current has no uniform distribution over the cross-section of the conductor because of the inner field in the conductor. The correction factor is, in that case, the ratio between the high-frequency inductance and the power frequency inductance of the high-voltage transmission line. Most high-voltage lines are equipped with *aluminium core steel reinforced* (ACSR) conductors, wherein the aluminium outer layer carries most of the current, and therefore the difference between the high-frequency inductance and the power frequency inductance is very small.

For the *last phase to clear* in the case of a three-phase fault, the characteristic impedances are connected as depicted in Figure 5.17 and phase $c$ is assumed to be the last clearing phase. The same connection of the characteristic impedances can be applied for a single-phase-to-ground fault if the two healthy phases are disconnected before the faulty phase. The sum of the line-to-ground $U_a$, $U_b$, and $U_c$ does not equal zero, and this indicates that both phase waves and earth waves exist. For a purely inductive line impedance the initial amplitudes will have, for the 50- or 60-Hz power frequency, the values

$$u_a = u_b = -1/3\ U_{\mathrm{phase}},\ u_c = 2/3\ U_{\mathrm{phase}},\ \text{and}\ u_{\mathrm{e}} = 1/3\ U_{\mathrm{phase}} \tag{5.21}$$

In phase $c$, both the phase wave and the earth wave are reflected at the short-circuit point. Till the reflected waves have travelled back, the breaker side of the line the voltage of phase $c$ will be constant. The rate of rise of the recovery voltage for the last clearing phase can be calculated from Figure 5.17.

$$\frac{\mathrm{d}u}{\mathrm{d}t} = \frac{\mathrm{d}i}{\mathrm{d}t}(Z + Z_{\mathrm{e}}) = \frac{\mathrm{d}i}{\mathrm{d}t}(2Z_1 + Z_0)/3 \tag{5.22}$$

The time to the first peak of the line-side TRV is in this case not simply twice the length of the high-voltage transmission line from breaker terminal to short-circuit point divided by the velocity of the travelling

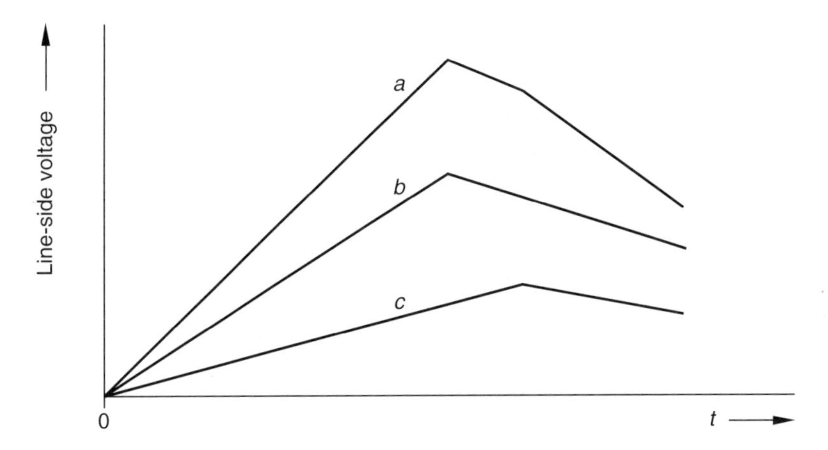


**Figure 5.18** The TRV waveform for the last phase to clear (a) is composed from a phase wave (b) and an earth wave (c)

waves. The phase wave and earth wave have a different velocity because they travel in a different medium and the earth wave is also distorted. The initial amplitude of the earth wave is only half the amplitude of that of the phase wave. Figure 5.18 depicts the influence of the earth wave on the TRV of the last phase to clear. The delay of the earth wave accounts for the distortion.

## 5.6 REFERENCES FOR FURTHER READING

Baatz, H., *Ueberspannungen in Energieversorgungsnetzen*, Part G, Chapters 6, 7, Springer-Verlag, Berlin, 1956.

Cigré SC 13, WG 05, "The calculation of switching surges," *Electra* **19**, (1971).

Cigré SC 13, WG 05, "The calculation of switching surges II: Network representation for energisation and re-energisation studies on lines fed by an inductive source" *Electra* **32**, (1974).

Cigré SC 13, WG 05, "The calculation of switching surges III: Transmission line representation for energisation studies with complex feeding networks, *Electra* **62**, (1979).

Dorsch, H., *Ueberspannungen und Isolationsbemessung bei Drehstrom Hochspannungsanlagen*, Chapter 1, Siemens Aktiengesellschaft, Berlin, 1981.

Ferschl, L. *et al.*, "Theoretical and experimental investigations of compressed gas circuit-breakers under short-line fault conditions," Cigré SC 13, Report of WG 07, 1974.

Garzon, R. D., *High-Voltage Circuit Breakers*, Chapter 3–5, Marcel Dekker, New York, 1997.

Greenwood, A., *Electrical Transients in Power Systems*, 2nd ed., Chapters 3, 5, Wiley & Sons, New York, 1991.

Happoldt, H. and Oeding, D., *Elektrische Kraftwerke und Netze*, Chapter 17, Springer-Verlag, Berlin, 1978.

Humphries, M. B. and Dubanton, M., "Transient Recovery voltage in the short line fault regime Initial Transient Recovery Voltage," Plenum press, NewYork, 1978.

Kennedy, M. W., in Flurscheim, C. H., ed., *Power Circuit Breaker: Theory and Design*, Chapter 3, Peter Peregrinus, London, 1975.

Nakanishi, K., ed., *Switching Phenomena in High-Voltage Circuit Breakers*, Chapter 1, Marcel Dekker, New York, 1991.

Peterson, H. A., *Transients in Power Systems*, Chapter 6, Wiley & Sons, New York, Chapman & Hall, London, 1951.

Ragaller, K., ed., *Current Interruption in High-Voltage Networks*, Plenum Press, New York, 1978.

Ruedenberg, R., *Elektrische Wanderwellen*, Part II, Chapter 9, Springer-Verlag, Berlin, 1962; Ruedenberg, R., *Electrical Shock Waves in Power Systems*, Part II, Chapter 9, Harvard University Press, Cambridge, Massachusetts, 1962.

Ruedenberg, R., in von H., Dorsch and P., Jacottet, eds., *Elektrische schaltvorgaenge*, 5th ed., Vol. II, Chapter 13, Springer-Verlag, Berlin, 1974.

Slamecka, E. and Waterschek, W., *Schaltvorgaenge in Hoch- und Niederspannungsnetzen*, Chapters 3, 4, Siemens Aktiengesellschaft, Berlin, 1978.

Thoren, B., "Short-Line Faults," *Elteknik* **9**(2), 25–33 (1966).

# 6

# Power System Transient Recovery Voltages

When interrupting devices, whether they are high-voltage circuit breakers, disconnectors, or fuses interrupt a current in the power system, the interrupting device experiences a recovery voltage across its terminals. When we bear in mind that during normal system operation the energy stored in the electromagnetic field is equally divided over the electric field and the magnetic field, current interruption causes a transfer of the energy content to the electric field only. This results in a voltage oscillation. The actual waveform of the voltage oscillation is determined by the configuration of the power system. The transient recovery voltage or TRV is present immediately after the interruption of the current. When the TRV oscillation has damped out, the power frequency–recovery voltage is active. The duration of the TRV is in the order of milliseconds, but its rate of rise and its amplitude are of vital importance for a successful operation of the interrupting device.

In the early years of switchgear and fuse design, the TRV was an unknown phenomenon, the recovery voltage was regarded to consist of the power frequency–recovery voltage only. The duty on circuit breakers was commonly expressed in terms of the circuit voltage prior to short circuit and the magnitude of the current in the arc. It was, however, experienced that, in practice, other circuit characteristics would affect the duty to an important extend [1]. Improved measurement equipment, such as the cathode-ray oscillograph and later the cathode-ray oscilloscope, made measurements with a higher time resolution possible and revealed the existence of a high-frequency oscillation immediately after current interruption: the TRV was discovered. This resulted in system studies of transmission and distribution networks. Many investigations were carried out in different countries to determine the TRV across circuit breakers while clearing short circuits in networks to provide a base for standardisation of the TRV in national rules for type tests on circuit breakers. System studies on transient network analysers brought insight into the frequencies of oscillations and were verified by real tests in the network [2]. Better understanding of the transient phenomena in the network has led to improved testing practice in the high-power laboratories, more accurate measurement of the current-zero phenomena, and consequently resulted in more reliable switchgear with a higher interrupting capability. The use of $SF_6$ as extinguishing medium allowed a leap forward in the short-circuit performance of high-voltage circuit breakers. But it appeared that until that time, unknown transients played an important role. It was not sufficient anymore to simulate the networks with lumped elements, and travelling waves also had to be taken into account. The short-line fault made its appearance [3], (see Chapter 5, *Switching Transients*). During that period of time, the end of the 1950s, effort was made to represent the TRV oscillations by standardised waveforms to be able to create TRVs by lumped elements within the walls of the high-power laboratory. The four-parameter waveform was proposed by Hochrainer in 1957 [4], and the possibilities to create the waveform in the high-power laboratory were investigated by Baltensperger [5].

In the early nineteen sixties, several network studies were undertaken in Japan [6] and Europe [7] and attempts were made to better-define the analogue modelling of the transient phenomena. The tests for TRVs were specified differently in the various national rules, and Subcommittee 17 on High-Voltage Switchgear of the International Electrotechnical Commission (IEC) asked the (Conférence Internationale des Grands Réseaux Electriques à Haute Tension) Study Committee on Circuit Breakers (named Study Committee 3 in those days) to promote new extensive investigations on an international base. A working group, CIGRE working group 3.1, set up for this purpose in 1959, decided to start a complete investigation of TRVs associated with the opening of the first pole of a circuit breaker clearing a three-phase ungrounded fault in some large 245 kV networks. Two of these networks were fully investigated: the Italian network in its 1962 situation and the French network in its 1965 situation. Some 2000 TRV waveshapes associated with short-circuit currents up to 45 kA were collected. Based on this collection of TRVs, a classification of current ratings was proposed [7].

In publication 56-2 (1971) on high-voltage alternating circuit breakers, IEC recommends characteristic values for simulation of the TRV by the four-parameter method ($U_1$, $t_1$, $U_c$, and $t_2$) or by the two-parameter method ($U_c$ and $t_3$). The values in the tables were mainly based on studies of the 245 kV systems. For the higher operating voltages up to 765 kV the values were extrapolated because actual data were hardly available at that time. In the meantime, studies were made of TRVs in systems operating at a maximum voltage of 420 kV and above. In some cases, these studies revealed deviations from the values on which the requirements stipulated in the IEC 56-2 (1971) were based.

In the light of this, CIGRE Study Committee 13 (Switching equipment) commissioned Working group 13.01 to study the problems of TRVs in extra-high-voltage systems [9]. One of the conclusions was, for example, that the rate of rise of 1 kV/μs, taken as a basis for ungrounded three-phase terminal faults, appeared to be somewhat low and that the stress to be expected was better characterised by a rate of rise of 2 kV/μs and a first-pole-to-clear-factor of 1.3.

It was decided in IEC-SC-17A to revise the TRV tables of the IEC 56-2 and IEC 56-4. New values were proposed based on studies, mentioned above, made by CIGRE Working group 13.01 during the years 1976–1979. Because these studies dealt mainly with rated voltages of 300 kV and above, and questionnaires issued to the utilities gave no indication about serious failures of medium-voltage breakers, IEC-SC-17A decided in 1979 only on the standard TRV values for rated voltages of 100 kV and above. The TRV values for rated voltages below 100 kV remained unchanged. CIGRE SC 13 decided at its meeting in Sydney in 1979 to set up a task force to collect data to establish TRV parameters for medium-voltage circuit breakers. The report of the task force was published in Electra No. 88 [10].

In June 1981, WG 13-05 was initiated by CIGRE-SC-13 to study the TRV conditions caused by clearing transformer and series reactor–limited faults. The Working Group collected data on transformer natural frequencies and ratings. The report of the working group was published in Electra No. 102 [11]. A combined CIGRE/CIRED working group WG CC-03 undertook from 1994 until 1998 the work of investigating the TRVs in networks till 100 kV. A summary of the report of WG CC-03 is published in Electra No. 181 [12]. Presently, each IEC standard has a code consisting of five numbers, for example, IEC 60056, IEC 60129, IEC 61633 and so on. The standard for high-voltage circuit breakers IEC 60056 (1987) is the fourth edition and will be, as the result of the work of Study Committee 17A WG 21, replaced by a fifth edition in 2001. For high-voltage circuit breakers also, the ANSI/IEEE standards are of importance. ANSI stands for the American National Standards Institute and IEEE stands for the

Institute of Electrical and Electronics Engineers. The most important ones are as follows:

| | |
|---|---|
| IEEE Std C37.04-1999 | IEEE Standard Rating Structure for AC High-Voltage Circuit Breakers |
| ANSI C37.06-1997 | AC High-Voltage Circuit Breakers Rated on a Symmetrical Current Basis - Preferred Ratings and Related Required Capabilities |
| IEEE Std C37.09-1999 | IEEE Standard Test Procedure for AC High-Voltage Circuit Breakers Rated on a Symmetrical Current Basis |

The IEC and ANSI/IEEE standard bodies are trying to harmonise their documents in the future.

## 6.1 *CHARACTERISTICS OF THE TRANSIENT RECOVERY VOLTAGE*

TRVs appear across the contacts of the interrupting device after every switching action. In switchgear standards, the main emphasis, however, is put on the interpretation of short-circuit or fault currents, because the stressing of circuit breakers is determined by the magnitude of the short circuit to be cleared and by the rate of rise and the peak value of the TRV. The IEC characterises the TRV by the four-parameter method ($U_1$, $t_1$, $U_c$, $t_2$) or by the two-parameter method ($U_c$, $t_3$). The IEC two-parameter and four-parameter limiting curves are depicted in Figure 6.1.

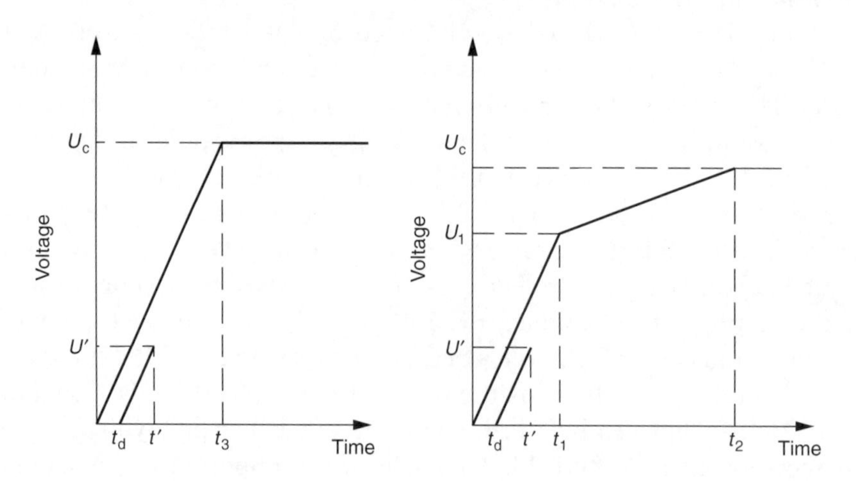


**Figure 6.1** IEC two- and four-parameter limiting curves

The two-parameter limiting curve is for high-voltage circuit breakers with a nominal rated voltage up to 100 kV and the four-parameter TRV applies to breakers with a rating of 100 kV and above. The parameter values for the TRV limiting curves are based on network studies in the 1960s conducted in the French and in the Italian grid [7]. In those days, the highest common transmission voltage was 245 kV. For higher transmission voltages, the network data were scarce and therefore the TRV parameters for voltage ratings up to 765 kV were extrapolated from the 245 kV data. Another limiting aspect is that only three-phase ungrounded faults at breaker terminals were considered and that only TRVs associated with the opening of the first pole of the breaker were determined. The three-phase ungrounded terminal fault was the base for circuit breaker testing in many national rules as well as in the IEC standard and the evaluation of the TRV for the first clearing pole is less complicated than for other types of faults, such as the single-phase-to-ground fault. The TRV parameters were grouped for certain current values, being 10%, 30%, 60%, and 100% of the maximum short-circuit current rating for the specific voltage level of 245 kV. In the IEC standards, these current values are referred to as duties, and the corresponding TRV parameters are tabled for these duties.

### 6.1.1 Short-Circuit Test Duties, Based on IEC 60056-1987

The TRV waveforms for voltage ratings up to 100 kV have a so-called (1-cosine) waveform because the interrupted short-circuit current is mainly supplied from substation transformers.

**10 percent short-circuit current; IEC test duty T10**

In the older version of the IEC-standard, this duty was named *test duty 1*.The fault current is supplied through one transformer only. The TRV has a (1-cosine) shape and a rather steep rate of rise, ranging from 5.5 kV/μs at 100 kV to 12.6 kV/μs at 765 kV. The current to be interrupted is symmetrical resulting in the highest peak voltage across the breaker terminals.

**30 percent short-circuit current: IEC test duty T30**

Former *test duty 2*. The fault current is 30 percent of the nominal rated short-circuit current and is supplied by a local source or one or two overhead lines, eventually with one or two transformers in parallel. The current might contain a considerable DC component with a time constant ranging from 150 till 300 milliseconds. The rate of rise of the TRV is lower than that for the duty T10, for instance, 5 kV/μs for breakers with a nominal voltage rating of 100 kV and above. To acquire the highest peak value for the TRV, the tests are carried out with a symmetrical current.

### 60 percent short-circuit current: IEC test duty T60

In the older versions of the IEC standard this duty was named *test duty 3*. The fault current is associated with 2, 3, or 4 overhead lines, eventually with transformers in parallel and the TRVs have a rate of rise of 3 kV/μs which is standardised for breakers with a voltage rating of 100 kV and above. The tests are also performed with a symmetrical current to get the highest TRV peak.

### 100 percent short-circuit current–IEC test duty T100s

Former *test duty 4*, the rate of rise of the TRV is 2 kV/μs, and by performing this test duty, the circuit breaker proves that it is able to interrupt its maximum rated symmetrical short-circuit current with its rated operating sequence.

There are two alternative rated operating sequences:

a) O–t–CO–t′–CO

Unless otherwise specified:

t = 3 min for circuit breakers not intended for rapid auto reclosing;
t = 0.3 s for circuit breakers intended for rapid auto reclosing (dead time);
t′ = 3 min.

(*Note:* Instead of t′ = 3 min, other values t′ = 15 s and t′ = 1 min are also used for circuit breakers intended for rapid auto reclosing.)

b) CO–t″– CO

With

t″ = 15 s for circuit breakers not intended for rapid auto reclosing;
O represents an opening operation;

**CO** represents a closing operation followed immediately (that is without an intentional delay) by an opening operation;
**t**, **t′** and **t″** are the time intervals between successive operations;
**t** and **t′** should always be expressed in minutes or in seconds;
**t″** should always be expressed in seconds.

The **O–0.3 s–CO–3 min–CO** duty cycle is typical for many high-voltage circuit breakers in service. The rapid auto reclosing, 300 ms after interruption of the fault current, is to verify whether the fault was created by touching lines or the current was injected in the system by a lightning stroke.

### 100 percent asymmetrical short-circuit current–IEC test duty T100a

For the majority of the circuit breakers, this is a very severe test because maximum arc energy is developed in the interrupting chamber of the breaker. Because of the maximum developed arcing energy, this is also a severe test for the operating mechanism. For gas-insulated metal-enclosed switchgear or GIS, the exhaust of gas can influence the capacity to withstand high voltage between phases and between phases and the tank. The time constant of the DC component is fixed at a standard value of 45 ms. Depending on the system characteristics, for example, if a circuit breaker is close to a generator, the time constant may be higher than the standard value. Recommended values are as follows:

- 120 ms for rated voltages up to and including 52 kV;
- 60 ms for rated voltages from 72.5 kV up to and including 420 kV;
- 75 ms for rated voltages 550 kV and above.

The basis for the IEC standards was developed in the 1960s and the ANSI/IEEE TRV parameters were also fixed in that period of time. While the requirements are similar, there are significant differences. ANSI C37.06 defines two TRV envelopes. For circuit breakers rated below 100 kV, Figure 6.2 shows the envelope defined as a (1-cosine) waveform with a crest value $E_2$ equal to 1.88 times the rated maximum voltage $V$ of the circuit breaker and the time to peak $T_2$ varies as a function of the voltage rating.

For circuit breakers rated above 100 kV, the envelope is defined in terms of an exponential-cosine waveform, as shown in Figure 6.3. The exponential is defined by an initial rate-of-rise parameter $R$, which has been established at 2 kV/μs, and a value $E_1$, which equals $1.3\sqrt{2}/\sqrt{3} = 1.06$

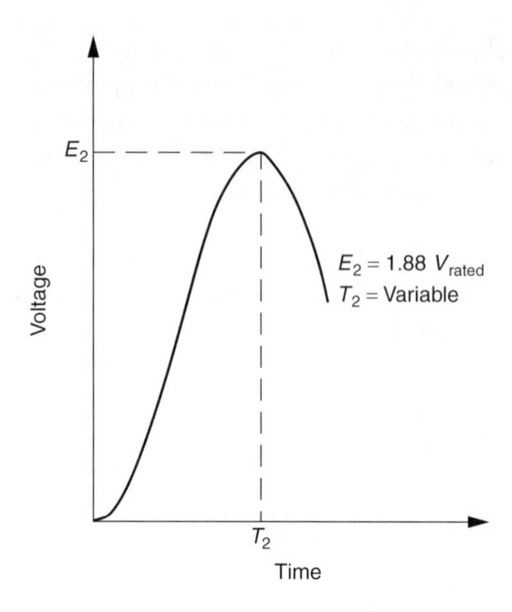


**Figure 6.2** IEEE standard TRV limiting curve for breakers rated below 100 kV

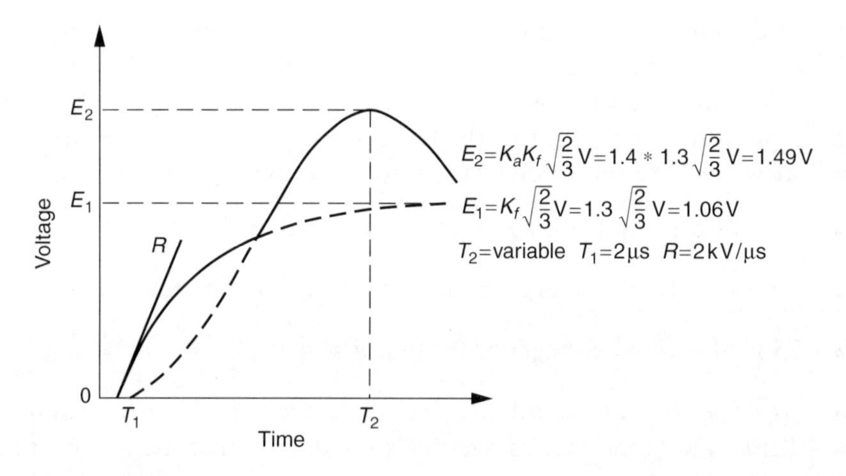


**Figure 6.3** IEEE standard TRV limiting curve for breakers rated 100 kV and above

times the rated maximum voltage. The (1-cosine) waveform is defined by a peak value of 1.49 times the rated maximum voltage and a time to peak $T_2$, which varies with the current and voltage rating of the breaker. A time-delay parameter $T_1$ in the initial part of the TRV is also specified to account for the effect of bus and breaker capacitance. The main differences between the IEEE and IEC standards with respect to the high-voltage TRV waveform are as follows:

- Below 100 kV, IEEE uses a (1-cosine) TRV waveform form while IEC uses a two-parameter envelope. For 100 kV and above voltage rating, IEEE makes use of an exponential-cosine waveform, whereas IEC uses a four-parameter envelope.
- The TRV parameters for the IEC envelopes are independent of the rated interrupting currents, whereas IEEE has different values for different ratings.
- IEC makes distinction between earthed neutral systems and isolated neutral systems by means of a 1.3 and 1.5 first-pole-to-clear factor, whereas IEEE fixes the first-pole-to-clear factor at 1.3 for all voltage ratings above 100 kV.
- The maximum peak factor $U_c$ of the TRV envelope of IEC is lower than the $E_2$ ANSI peak value for breaker ratings below 100 kV.

### 6.1.2 Short-Circuit Test Duties Based on ANSI/IEEE Standards

The test duties resemble the test duties of the IEC 60056. For voltages below 100 kV, the TRV waveform has a (1-cosine) shape. The time to peak $T_2$ corresponds with 1.137 times $t_3$ (or $t_3 = 0.88T_2$) from the IEC. For voltages below 100 kV, ANSI/IEEE assumes an ungrounded system by means of a first-pole-to-clear factor of 1.5, and the peak value of the TRV is $E_2 = 1.88$ kV. Expressed in terms of an amplitude factor, this corresponds to an amplitude factor of $Ka = 1.535$, whereas the IEC requires an amplitude factor of 1.4 related with a 100 percent short-circuit current.

For voltage ratings of 100 kV and above, ANSI/IEEE is also in close harmony with IEC. For the TRV, the exponential-cosine waveform is used but the rate-of-rise and the time-delay are as in IEC: 2 kV/μs, and 2 μs, respectively. Further, IEEE Std C37.04-1999 reads: 'Since most, if not all, systems operating at 100 kV and above are effectively grounded, a first-pole-to-clear factor of 1.3 is required.' At this point, ANSI/IEEE differs from IEC. For voltage ratings between 100 kV up to and including 170 kV, IEC offers the choice between an isolated neutral, and an earthed neutral, and depending on the system grounding the first-pole-to-clear factor is 1.5 and 1.3, respectively.

The amplitude factor of the TRV oscillation is, similar to IEC, 1.4, and this corresponds with the parameters of the exponential cosine: $E_1 = 1.3 * \sqrt{2}/\sqrt{3}$ V $= 1.06$ V and $E_2 = 1.3 * 1.4 * \sqrt{2}/\sqrt{3}$ V $= 1.49$ V.

### 6.1.3 The Harmonisation between IEC and ANSI/IEEE Standards with respect to Short-Circuit Test Duties

A combined IEC and ANSI/IEEE working group, WG 23, pursues more harmonisation between IEC and ANSI/IEEE standards dealing with short-circuit test duties. Both standards will undergo changes and the goal is to come to one standard for high-voltage circuit breakers. The most likely outcome is that for circuit breakers with a rating below 100 kV the TRV will be described by a two-parameter waveform and the values for $U_c$, $t_3$, and $t_d$ are the values from IEC. For circuit breaker ratings of 100 kV and above, the IEEE adopts the four-parameter waveshape of IEC, instead of its exponential-cosine waveform, with a value for $U_1$ reduced to $0.75 * U_1$ of the current value prescribed by IEC 60056. The test duties T10 (formerly *duty* 1) and T30 (formerly *duty* 2) will have a two-parameter waveshape, and for test duty T30 the amplitude factor is raised from 1.5 till 1.54.

## 6.2 *THE TRANSIENT RECOVERY VOLTAGE FOR DIFFERENT TYPES OF FAULTS*

The TRV, which stresses the circuit breaker after current interruption, depends on the type of fault, the location of the fault and the type of circuit switched. To determine the TRV after current interruption, we can make use of the superposition principle: when an identical current of opposite polarity is superimposed on the short-circuit current, the original short-circuit current is interrupted. The TRV can be determined by injecting a corresponding current across the opening breaker poles in the network in which the voltage sources are short-circuited and the current sources, if any, are removed. Figure 6.4 shows the three-phase-to-ground fault supplied from a delta-wye transformer. This fault has already been introduced in Chapter 2, *The Transient Analysis of Three Phase Power Systems*, where the neutral of the supply was grounded via an impedance and the three-phase-to-ground fault was treated with fault impedance $Z_f$ and $Z_g$. The connection of the sequence networks in Figure 6.4b resembles Figure 2.9 with $Z_f = Z_g = 0$.

The steady-state voltage on the fault side of the first opening pole of the breaker is zero, whereas on the supply side, the value of the steady-state voltage is 1.0 per unit. In other words, the first-pole-to-clear factor is 1.0 because the grounded supply *and* the three phase grounded fault make it a symmetrical system. The supplying transformer is not only a pure short-circuit reactance but also has a capacitance between the windings and a

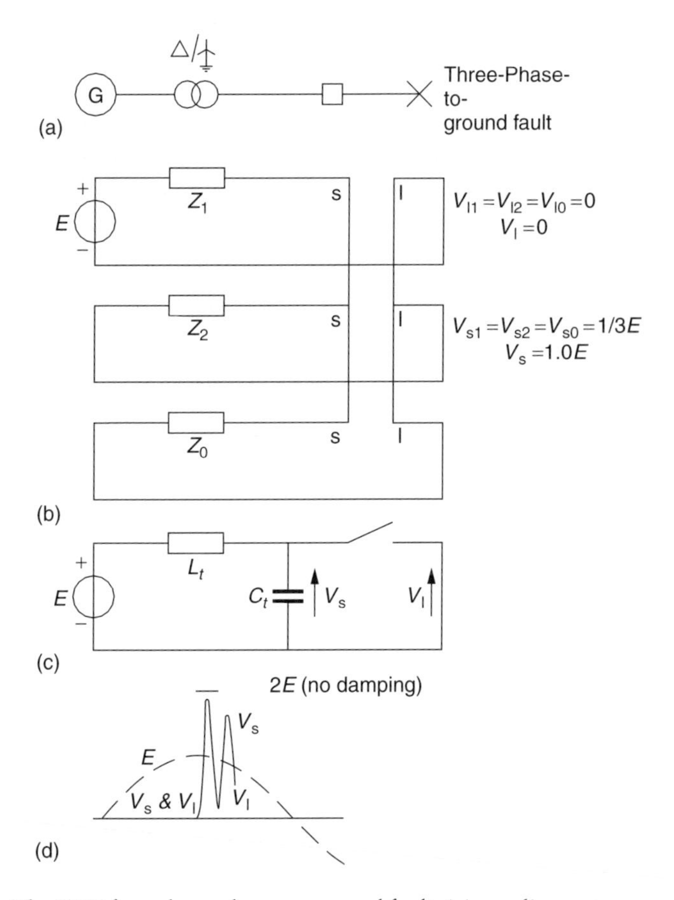


**Figure 6.4** The TRV for a three-phase-to-ground fault: (a) one-line system representation; (b) connection of the sequence networks; (c) transient network; (d) system voltages

capacitance to ground. This distributed capacitance can be represented as a lumped capacitance as shown in Figure 6.4c. Further, it is assumed that the supplying synchronous generators are *electrically* far away, and the influence of the sequence impedance of the generators on the sequence networks can be neglected. When the three breaker poles are still arcing, the voltages on both sides of the breaker are equal to zero. Immediately after the first pole clears the current, $V_l$ remains equal to zero but $V_s$ increases to $E$, which is 1.0 per unit. With no damping in the circuit, $V_s$ would reach a peak value of 2.0 per unit, because the transient voltage oscillation is a (1-cosine) waveform as illustrated in Chapter 1, *Basic Concepts and Simple Switching Transients*, but with damping (which is always present in a practical situation due to iron and copper losses), the peak voltage would be somewhat less than 2.0 per unit. A typical value is a peak value of 1.8 per unit. The frequency of the transient voltage oscillation is a function of the inductance and the lumped capacitance of the transformer.

Figure 6.5 shows a similar calculation for an ungrounded three-phase fault. When the first-pole-to-clear has interrupted the current, the three-phase system is not symmetrical anymore. At the supply side of the breaker, the voltage is 1.0 per unit, but at the fault side, the voltage is −0.5 per unit (see Chapter 2, *The Transient Analysis of Three-Phase Power Systems*). $V_s$ starts at zero and oscillates around $E$, with a frequency determined by the transformer inductance and capacitance and would reach without damping a peak value of 2.0 per unit. $V_l$ at the load side of the breaker starts at zero and oscillates around −0.5 per unit at the same frequency.

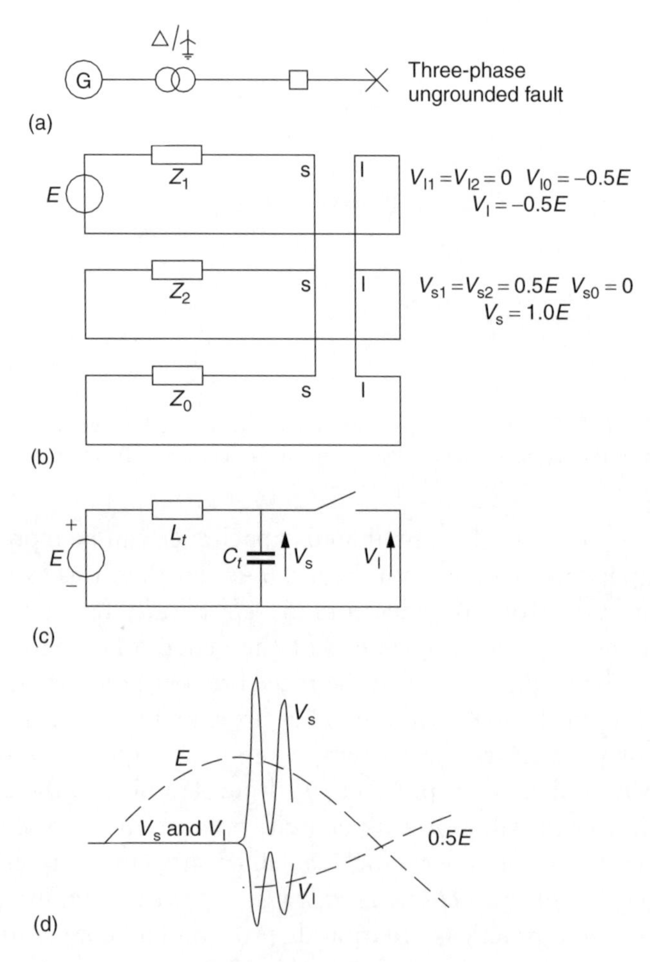


**Figure 6.5** The TRV for a three-phase-ungrounded fault: (a) one-line system representation; (b) connection of the sequence networks; (c) transient network; (d) system voltages

The TRV across the breaker terminals is $V_s - V_l$ and reaches, with no damping, a peak value of 3.0 per unit.

If the neutral of the supply transformer in Figure 6.4a is ungrounded, the zero-sequence source circuit is open, so that $V_{s1}$, $V_{s2}$, and $V_{s0}$ in Figure 6.4b are all equal to 0.5 $E$. The steady-state value of the voltage at the supply side of the circuit breaker is in that case 1.5 per unit. The value of the steady-state voltage at the fault side of the breaker is zero and so the voltage across the breaker terminals is again 1.5 per unit, and the TRV again an oscillatory wave with a peak vale of 3.0 per unit assuming no damping. Thus a three-phase grounded fault in an ungrounded system results in the same TRV as a three-phase ungrounded fault in a grounded system. In both situations, the first-pole-to-clear factor is 1.5.

These circuits, in which the fault current is supplied from a single source, show the basics behind the TRV oscillations and the influence of system grounding. The circuit layout is rather simple, but in practice (especially at the higher system voltages), the configuration is more complex. The source of the fault current is not only a local transformer but a considerable part of the current is supplied from other sources further away over transmission lines. This results, after interruption of the fault current, in a TRV, which is composed of a local oscillation from the lumped transformer inductance and capacitance, and from reflecting and refracting travelling waves from the transmission lines. (See Section 3.7, *The Origin of Transient Recovery voltages.*) In practice, the waveshape is therefore different for each fault situation and can have, in some cases, a rather unpredictable shape. The testing standards can obviously not cope with this and therefore TRV waveforms are standardised with two-parameter and four-parameter envelopes.

## 6.3 REFERENCES


6.1 Park, R. H. and Skeats, W. F., "Circuit breaker recovery voltages, Magnitudes and Rates of Rise", *Trans. A.I.E.E.*, 204–239 (1931).

6.2 Catenacci, G., "Le frequenze proprie della rete Edison ad A.T.," *L'Elettrotrechnica* **XLIII**(3), (1956).

6.3 Pouard, M., "Nouvelles notions sur les vitesses de rétablissement de la tension aux bornes de disjoncteurs à haute tension," *Bulletin de la Société Française des Electriciens*, 7th series **VIII**(95), 748–764 (1958).

6.4 von Hochrainer, A., "Das Vier-Parameter-Verfahren zur Kennzeichnung der Einschwingspannung in Netzen" *ETZ* **78**(Part 19), 689–693 (1957).

6.5 Baltensperger, P., "Définition de la tension transitoire de rétablissement aux bornes d'un disjoncteur par quatre paramètres, possibilités des stations d'essais de court-circuit," *Bulletin de l'Association Suisse des Electriciens* 3, (1960).

6.6 Ozaki, Y., "Switching surges on high-voltage systems," Central Research Institute of Electric Power Industry, Tokyo, Japan, 1994.
6.7 Catenacci, G., Paris, L., Couvreux, J. P., and Pouard, M., "Transient recovery voltages in French and Italian high-voltage networks," *IEEE*, **PAS 1986**(11), 1420–1431 (1967).
6.8 Barret, J. P., "Développements récents des méthodes d'étude des tensions transistoires de manoeuvre sur les réseaux à haute tension, " *Revue générale de l'électricité*, 441–470 (1965).
6.9 Braun, A. *et al.*, "Characteristic values of the transient recovery voltage for different types of short circuits in an extensive 420 KV system," *ETZ-A*, **97**, 489–493 (1976).
6.10 Novotry, V. *et al.*, "Transient recovery voltages in medium-voltage networks," *Electra* **88**, 49–88 (1983).
6.11 Parrot, P. G., "A review of transformer TRV conditions," *Electra* **102**, 87–118 (1985).
6.12 Bonfanti, I. *et al.*, "Transient recovery voltages in medium-voltage networks," *Electra* **181**, 139–151 (1998).

# 7

# Lightning-Induced Transients

Overvoltages in a power system can be caused by transient currents and by transient voltages after switching actions during normal operation or after clearing fault situations. The overvoltages originate from the state of the system. There are also overvoltages that come from outside the system as a result of atmospheric discharges. Large parts of the power system are formed by overhead transmission lines interconnected by outdoor substations. Only in densely populated areas, the high-voltage transmission and distribution is done with high-voltage cables interconnected by gas-insulated substations (GIS) placed in buildings. When we realise that on an average every commercial aeroplane and every square kilometre of the earth's surface in a country like the Netherlands is hit by lightning once a year, it is obvious that our power systems should be protected against lightning strokes. For the analysis of the lightning-induced overvoltages, a difference is made between the following:

- Lightning strokes in the vicinity of high-voltage transmission lines, which do not hit the conductors themselves,
- Direct lightning strokes on the line conductors injecting a current wave on the line, and
- Lightning strokes on the transmission towers or on the ground wires.

In addition, the solar system can be the cause of a blackout of the power system. Heavy eruptions on the solar surface in a so-called solar spot on March 10th, 1989, caused a large blackout of the Hydro Quebec grid in Canada on March 13th. Geomagnetic-induced currents were the cause of a DC component in the system current, and the power transformers were brought into saturation. The system current got strongly distorted and contained, apart from the power frequency ground wave, higher harmonics. The higher harmonics increased the apparent reactance of the power transformers, and therefore the transformer also consumed more reactive power. The temperature of the transformer went up, and after a certain period of time, the transformer protection tripped the power transformer out of service.

## 7.1 THE MECHANISM OF LIGHTNING

Lightning mostly occurs on summer days when the ambient temperature is high and the air is humid. Because of the temperature difference, the humid air is lifted to higher altitudes with a considerable lower ambient temperature. Cold air contains less water than warm air and raindrops are formed. The raindrops have a size of a few millimetres and are polarised by the electric field that is present between the lower part of the ionosphere and the earth's surface. The strength of this atmospheric field is on summer days in the order of 60 V/m and can reach values of 500 V/m on a dry winter day. The vertical movement of the raindrops and the wind shear splits the raindrops into negatively charged small drops and positively charged larger raindrops. The larger raindrops fall, under the influence of the gravity, to earth and create a shower. The positive charge of raindrops in a shower is confirmed by measurements. The majority of the thunderclouds is *negatively* charged with a potential to earth of several hundreds of megaVolts. The clouds move at great heights and the average field strength is far below the average breakdown strength of air. Inside the thundercloud, the space charge formed by the accumulating negative raindrops creates a locally strong electric field in the order of 10 000 V/m and accelerates the quickly moving negative ions to considerable velocities. Collision between the accelerated negative ions and air molecules, creates new negative ions, which on their part are accelerated, collide with air molecules and free fresh negative ions. An avalanche takes place, and the space charge and the resulting electric field grow in a very short period of time. The strong electric field initiates discharges inside the cloud, and a negative stream of electrons emerges as a dim spark called a *stepped leader* or *dart leader* that jumps in steps of approximately 30 meters and reaches the earth in about 10 milliseconds. The stepped leader reaches close to the earth's surface, reaching an upward positive leader, and forms the main channel.

Because of the stochastic behaviour of the space charge accumulation, the stepped leader also creates branches to the main channel. The main channel carries initially a discharge current of a few hundred Amperes, having a speed of approximately 150 km/s. This discharge current heats up the main channel and the main discharge, the *return stroke*, is a positive discharge, and it travels at approximately half the speed of light equalising the charge difference between the thundercloud and the earth. The main discharge current can be 100 000 A or more and the temperature of the plasma in the main channel can reach values as high as 30 000 K. The pressure in the main channel is typically 20 bar. The creation of the return stroke takes place between 5 and 10 μs and is accompanied by a shock wave that we experience as *thunder*. A lightning stroke consists of several of these discharges, usually three or four, with an interval time of 10–100 μs. The human eye records this as *flickering* of the lightning and our ear hears the *rolling* of the thunder. After each discharge, the plasma channel cools down to approximately 3000 K, leaving enough ionisation to create a new conducting plasma channel for the following discharge. In the majority of the cases, a thundercloud is negatively charged, but positively charged clouds can also be formed.

Cloud-to-ground strokes can hit substations, transmission line towers, and transmission lines directly, but a considerable number of atmospheric discharges are between clouds. When the charged clouds are floating above, for instance, a high-voltage transmission line, they induce charge accumulation on the line conductors (see Figure 7.1). When the lightning stroke equalises the charge difference between the clouds, the rather slowly accumulated charge on the conductors has to disappear at once. This results in transient currents and overvoltages.

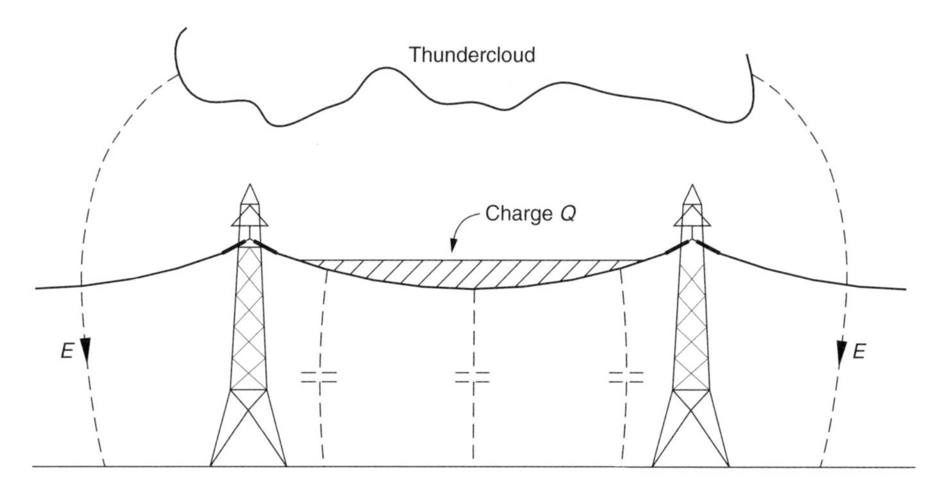


**Figure 7.1** Effect of a charged cloud on a high-voltage transmission line

## 7.2 *WAVESHAPE OF THE LIGHTNING CURRENT*

Lightning currents differ in amplitude and shape. The majority of the cloud-to-ground lightning strokes vary from kiloAmperes to several tenths of kiloamperes. Strokes above 100 000 amperes are rare, and the highest reported peak value of the return stroke current is 200 000 A. The shape of the current wave and the related voltage wave is rather capricious and different for every stroke. To facilitate testing in the laboratory and computations either by hand or by computer, the shape of the current wave of the return stroke is standardised. The IEC has standardised the so-called 1.2/50-μs waveform (Figure 7.2). The rise time $t_f = 1.2$ μs is defined as being 1.67 times the time interval between 30 percent and 90 percent of the peak value of the current wave. The tail value $t_t = 50$ μs is defined as the time it takes until the wave drops till 50 percent of the peak value. When the current wave travels in the power system, there is, of course, a related voltage wave also present. The ratio between the voltage wave and the current wave at a certain place in the system is the characteristic impedance at that particular part of the network. System components can be exposed to very high lightning-induced overvoltages. The name plate of high-voltage equipment shows the *Basic Insulation Level or BIL*, which is a standardised figure for each voltage rating related to the voltage level at which the equipment should operate.

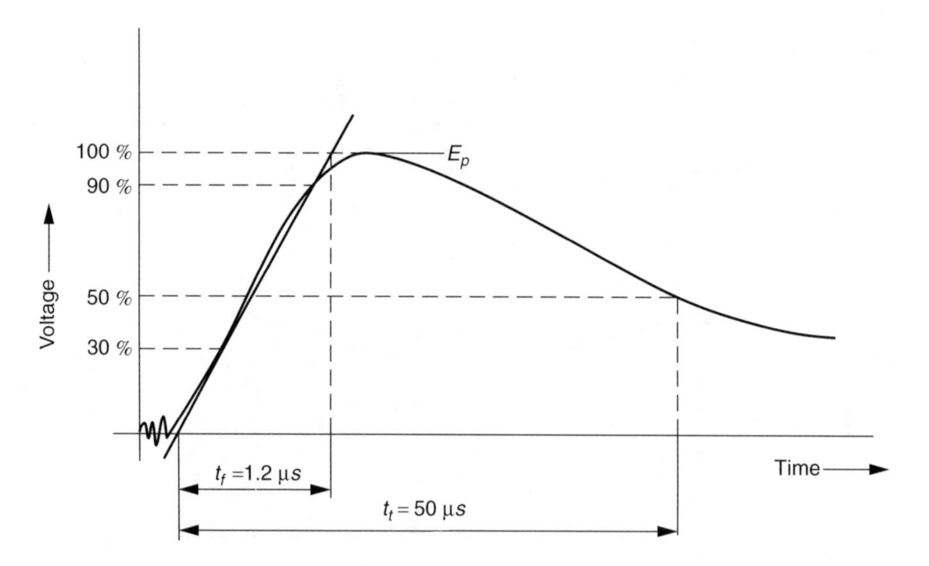


**Figure 7.2** Standardised waveform of a lightning-induced voltage wave

The lightning-induced voltage wave can be described mathematically as the difference of two exponential functions:

$$e(t) = E_p(e^{-\alpha t} - e^{-\beta t}) \tag{7.1}$$

In this expression, the parameter $\beta$ is associated with the rise time $t_f$ and $\alpha$ with the tail time $t_t$. When choosing for $\alpha$ the value $1.4 * E4\,s^{-1}$ and for $\beta$ the value $4.5 * E6\ s^{-1}$, the double exponential expression of Equation (7.1) results in a 1.2/50-microsecond waveform. The double exponential wave is easy to manipulate in mathematical analysis and results in an acceptable degree of accuracy.

## *7.3 DIRECT LIGHTNING STROKE TO TRANSMISSION LINE TOWERS*

The probability of a direct hit on a transmission line tower is high compared with the vulnerability to lightning of other parts of the power system. Transmission line towers are interconnected by *ground* wires, installed to reduce the voltages induced on the phase conductors by lightning strokes to nearby ground and to shield phase conductors from direct lightning strokes. Ground wires (also named *shield wires*) hang above the phase conductors in the tower and are electrically connected with the tower frame and through the towers to the ground below. The phase conductors are electrically isolated from the towers by means of insulator strings to withstand the power frequency system voltage and the transient voltages created by switching actions and lightning.

The ground wires form a shield for the phase wires against a direct hit by lightning and reduce the tower ground resistance in dry or rocky soil significantly. If a stepped leader approaches the earth, a positive leader reaches up from the nearest sharp conductor. When the leader arrives within striking distance of a grounded object, which can be a tree, a church tower, or an animal or human being in the field, flashover takes place. The shield wire is positioned such that the stepped leader makes contact with the shield wire instead of the phase conductors.

What peak voltages can we expect after a transmission line tower is hit by lightning? The instant the stepped leader reaches the tower, the thundercloud having a potential of $E$ Volts has a conducting path to the ground. The charge difference between the thundercloud and the earth is equalised by a travelling wave, moving upwards at approximately half the velocity of light. The discharge takes place in about 50 μs, the branches are brightly illuminated, and the amplitude of the current is in the range of 20–100 kA. The characteristic impedance of the lightning current path lies in the order of the characteristic impedance of overhead transmission lines $Z_{\text{L}} = 500\ \Omega$. The stroke current $I_{\text{stroke}}$ will divide into three parts. One part of the stroke current, $I_{\text{tower}}$, will flow through the tower structure to ground and the remaining part of the stroke current divides equally and flows in opposite directions through the ground wire, as shown in Figure 7.3.

$$I_{\text{stroke}} = I_{\text{tower}} + I_{\text{ground wire}} \tag{7.2}$$

After the stroke, the tower top has an initial voltage equal to the difference between the potential $E$ of the thundercloud and the voltage drop over the lightning path $V_L$. Three voltage waves, all equal to the initial tower top voltage, $V_{\text{tower}} = E - V_L$, travel away from the struck point – one voltage wave along the tower structure toward the ground and the other two in opposite directions from the struck point along the ground wire. The voltage waves travel with a velocity less than that of light in free space, and each of the three waves will be refracted and reflected at the nearest transition points. In the tower structure, there will be reflections between the tower-footing resistance and the tower top. The voltage waves travelling on the ground wires will be reflected and further transmitted to the adjacent towers. This process repeats itself as the transmitted voltage waves travel along the ground wire.

On the ground wire, a forward and backward current and the therewith-related voltage waves travel. The waves in the positive $x$-direction are the

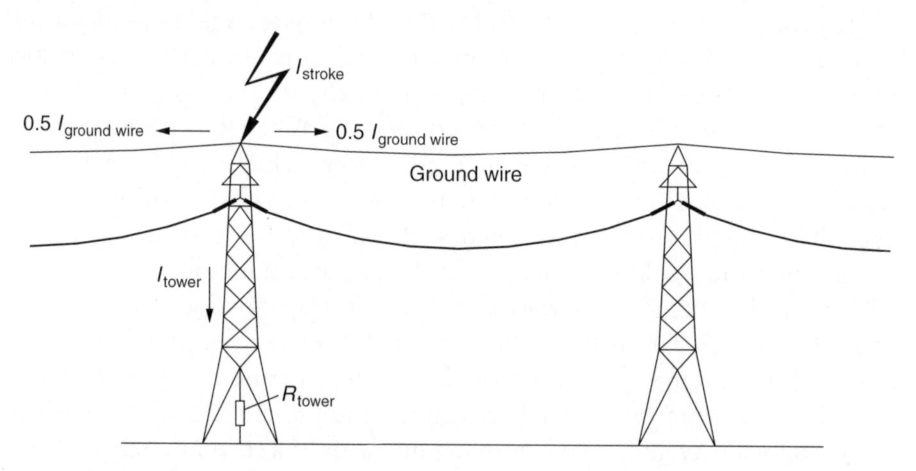


**Figure 7.3** Lightning stroke to a transmission line tower with ground wires

forward waves and the waves in the negative $x$-direction form the backward waves. This implies a minus sign for the backward-travelling current wave, because it propagates in the negative $x$-direction (see Chapter 3, *Travelling Waves*).

For the current in the ground wire, $I_{\text{forward}} = -I_{\text{backward}} = 0.5I_{\text{ground wire}}$, and for the voltage on the ground wire $V_{\text{tower}} = 0.5I_{\text{ground wire}} Z_{\text{ground wire}}$

Substituted in Equation (7.2)

$$I_{\text{stroke}} = \frac{V_{\text{tower}}}{R_{\text{tower}}} + \frac{2V_{\text{tower}}}{Z_{\text{ground wire}}} \tag{7.3}$$

The voltage at the tower top can be calculated with

$$V_{\text{tower}} = I_{\text{stroke}} \left\{ \frac{Z_{\text{ground wire}}}{2 + \dfrac{Z_{\text{ground wire}}}{R_{\text{tower}}}} \right\} \tag{7.4}$$

The voltage and current waves have the same shape as the lightning wave.

## *7.4 DIRECT LIGHTNING STROKE TO A LINE*

When lightning strikes directly on a phase conductor of a transmission line, it can be regarded as a *current injection I* on the line, which divides itself into two equal parts at the point of strike. The voltages generated by the divided currents travel in both directions along the line, away from the point of strike. When the characteristic impedance of the phase conductor is $Z_{\text{phase}}$, the voltage is related to the lightning current $I$ by

$$V = 0.5IZ_{\text{phase}} \tag{7.5}$$

The voltage wave travelling along the line hit by lightning, for instance, phase a, induces voltage waves on the neighbouring phases also. When the *coupling factor* (see Chapter 5, Section 5.1, *Interrupting of Capacitive Currents*) between phase a and phase b is $K_{ab}$ then the voltage to the ground, induced on phase b is

$$V_b = K_{ab} V_a \tag{7.6}$$

and the voltage between phase a and b is

$$V_{ab} = (1 - K_{ab})V_a \tag{7.7}$$

The voltage $V_{ab}$ can cause a flashover between the two phases, if the separation distance is not large enough, thus introducing a two-phase short-circuit. When no flashover occurs, $V_a$, $V_b$, and $V_c$ will travel along the phase conductors $a$, $b$, and $c$ and encounter the insulator strings that support the phase conductors in the tower, which is at ground potential. The voltage $V_a$ is the highest, and the insulator string of phase $a$ can be subjected to a flashover.

The voltage waves travel in both directions along the line, away from the point of strike, and they travel from tower to tower. In the vicinity of each tower, the phase conductor has a larger capacitance to ground than at other parts of the line, because of the insulator strings and the presence of the metal tower structure that brings the ground potential closer. This tower capacitance reduces the front steepness of the voltage wave and creates eventually a time delay. After passing a few transmission line towers, the danger of a flashover reduces considerably.

Travelling along the line, the voltage waves can also encounter a series capacitor for compensation of long transmission lines, a series reactor used for telecommunication over the line, or the transmission line connected to a high-voltage cable before entering a substation. In all cases, there is a change in characteristic impedance at the transition point and the relation between the current and voltage wave changes (see Chapter 3, Section 3.5, *Reflection and Refraction of Travelling Waves*). Examining the series connection of a high-voltage transmission line, an impedance, and a high-voltage cable, as depicted in Figure 7.4, gives already a clear insight into the voltage profile, arising after the reflection and refraction of the travelling voltage waves. Let us assume that voltage wave $e(t)$ is present on the high-voltage transmission line and travels toward A, where the series impedance is connected. For the ease of calculation, the voltage function and the lumped and distributed network elements are transformed to the Laplace domain. The lightning-induced voltage wave can, for a first-approximation, be described by a single exponential

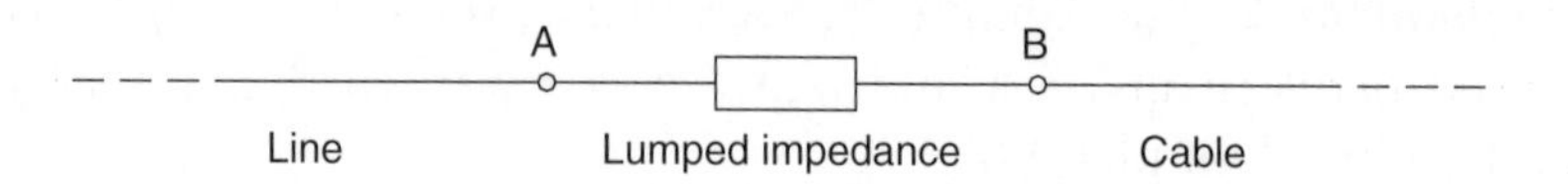


**Figure 7.4** Series connection of a high-voltage transmission line, an impedance, and a high-voltage cable

function

$$e(t) = E_p \mathrm{e}^{-\alpha t} \tag{7.8}$$

and as series impedance, we take an air-core reactor with resistive losses in the Laplace domain expressed as $Z(p) = pL + R$.

When, at the instant $t = 0$, the voltage wave $e(p)$ reaches point A, the voltage wave splits in a part that is reflected and travels backwards and a part that is refracted, which continues its way to point B.

The reflection coefficient for point A is

$$r_{\mathrm{A}} = \frac{Z(p) + Z_2 - Z_1}{Z(p) + Z_2 + Z_1} \tag{7.9}$$

The expression for the voltage at point A is

$$e_{\mathrm{A}}(p) = e(p)\left\{1 + \frac{Z(p) + Z_2 - Z_1}{Z(p) + Z_2 + Z_1}\right\} = 2e(p)\frac{Z(p) + Z_2}{Z(p) + Z_2 + Z_1} \tag{7.10}$$

The expression for the current wave travelling through $Z(p)$, and subsequently through $Z_2$ is

$$i_{\mathrm{B}}(p) = \frac{e_{\mathrm{A}}(p)}{Z(p) + Z_2} = \frac{2e(p)}{Z(p) + Z_2 + Z_1} \tag{7.11}$$

For the voltage at point B, we can write

$$e_{\mathrm{B}}(p) = 2e(p)\frac{Z_2}{Z(p) + Z_2 + Z_1} \tag{7.12}$$

Substituting for the lightning wave, the simplified expression of the single exponential function of Equation (7.8), which transformed to the Laplace domain, becomes $e(p) = E_p/(p + \alpha)$, and for the series impedance between the line and the cable, an air-core reactor with resistive losses, transformed to the Laplace domain $Z(p) = pL + R$, gives for the current through the air-core reactor

$$i_{\mathrm{B}}(p) = 2E_p \frac{1}{(p + \alpha)} \frac{1}{(pL + R + Z_1 + Z_2)} \tag{7.13}$$

and transformed back into the timed domain, the expression for the current is

$$i_{\mathrm{B}}(t) = 2E_p \frac{1}{Z_1 + Z_2 + R - \alpha L}\{\mathrm{e}^{-\alpha t} - \mathrm{e}^{-[(Z_1+Z_2+R)/L]t}\} \tag{7.14}$$

The two exponential terms in Equation (7.14) can be reduced to a single exponential term when $\alpha = 0$, and thus the travelling wave is in fact a step function, $e(p) = E_p/p$, and Equation (7.14) becomes

$$i_{\mathrm{B}}(t) = \frac{e_{\mathrm{B}}(t)}{Z_2} = 2E_p \frac{1}{Z_1 + Z_2 + R}\{1 - \mathrm{e}^{-[(Z_1+Z_2+R)/L]t}\} \tag{7.15}$$

The expression for the voltage at point A (see Figure 7.4) can be calculated with $e_{\mathrm{A}}(p) = i_{\mathrm{B}}(p)\{pL + R + Z_2\}$, and is transformed into the time domain,

$$e_{\mathrm{A}}(t) = \left\{E_p \frac{2(R + Z_2)}{Z_1 + Z_2 + R} + E_p \frac{2Z_1}{Z_1 + Z_2 + R}\mathrm{e}^{-[(Z_1+Z_2+R)/L]t}\right\} \tag{7.16}$$

Let us now examine a few possible configurations. First: the situation that line and cable are coupled directly. This means that in our example the values for $R$ and $L$ are zero. The voltage at the line–cable intersection is

$$e_{\mathrm{B}}(t) = E_p \frac{2Z_2}{Z_1 + Z_2} \tag{7.17}$$

The nominator is smaller than the denominator when $Z_1 > Z_2$, and the refracted wave, travelling from intersection point B onwards, in that case, is smaller in amplitude than the incident wave. When we consider 450 Ω as characteristic impedance for the transmission line and 50 Ω for the cable, the reduction in amplitude is eighty percent. In other words, when an incoming high-voltage transmission line is connected to a substation via an underground cable, the cable serves as an amplitude reductor for incoming lightning-impulse and switching-impulse waves. When the substation itself is hit by lightning, the opposite occurs; the wave travelling via the cable to the line is amplified at the line–cable intersection point, in our example by a factor 1.8.

When the connection between the line and the cable is formed by a component with a large inductivity, for instance, a current transformer, the series winding of a phase-shift transformer or a line-carrier reactor, the expression for the voltage at point B is, when we neglect resistive losses,

$$e_B(t) = E_p \frac{2Z_2}{Z_1 + Z_2}\{1 - e^{-[(Z_1+Z_2)/L]t}\} \tag{7.18}$$

and the voltage at point A

$$e_A(t) = \left\{E_p \frac{2Z_2}{Z_1 + Z_2} + E_p \frac{2Z_1}{Z_1 + Z_2} e^{-[(Z_1+Z_2)/L]t}\right\} \tag{7.19}$$

The voltage waves $E_A(t)$ and $E_B(t)$ are depicted in Figure 7.5 for the case $Z_1 = Z_2$

A series reactance between two lengths of a transmission line reduces the wave front of the refracted wave, but for the incoming wave, the reactance behaves first as an open connection, which results in a reflected wave with a double amplitude. In addition, the reactance itself experiences severe voltage stress.

The expression for the voltage $e_A(t) - e_B(t)$ across the reactance is

$$e_L(t) = e_A(t) - e_B(t) = \{2E_p e^{-[(Z_1+Z_2)/L]t}\} \tag{7.20}$$

From Equation (7.20), we can see that the reactance must be able to withstand twice the crest value of the incident lightning impulse wave or should be protected by surge arresters. The highest voltage appears across the terminals of the series reactance when the lightning wave reaches the reactance at $t = 0$. After a time duration of more than three reactance time constants, $t > 3L/R$, the influence of the reactance on the waveshape disappears and the situation is similar to direct coupled lines (Equation (7.17)).

When two transmission lines have a *parallel impedance* connected at their point of intersection, as depicted in Figure 7.6, the parallel impedance

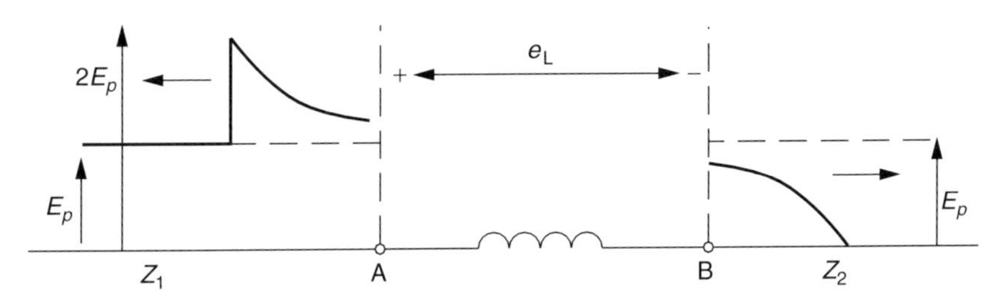


**Figure 7.5** Reflected and refracted travelling waves along two series connected lines and a series reactance

changes the characteristic impedance at the intersection point A, and for a lightning impulse wave, travelling from left to right, the coefficient of reflection at point A for a general parallel impedance is

$$r_{\mathrm{A}} = \frac{[Z(p)Z_2/(Z(p)+Z_2)] - Z_1}{[Z(p)Z_2/(Z(p)+Z_2)] + Z_1} = \frac{Z(p)(Z_2 - Z_1) - Z_1Z_2}{Z(p)(Z_2 + Z_1) + Z_1Z_2} \tag{7.21}$$

The voltage at point A is the addition of the incoming and the reflected wave

$$e_{\mathrm{A}}(p) = e(p)\{1 + r_{\mathrm{A}}\} = e(p)\frac{2Z_2Z(p)}{Z(p)(Z_2 + Z_1) + Z_1Z_2} \tag{7.22}$$

When the parallel impedance is a lumped capacitance, we can substitute

$$Z(p) = \frac{1}{pC} \tag{7.23}$$

Equation (7.22), the expression of the voltage at node A, becomes

$$e_{\mathrm{A}}(p) = e(p)\frac{2Z_2}{(Z_1 + Z_2) + pCZ_1Z_2} = e(p)\frac{2Z_2}{CZ_1Z_2}\frac{1}{p + \dfrac{(Z_1 + Z_2)}{CZ_1Z_2}} \tag{7.24}$$

As a first-approximation, the lightning impulse wave can be regarded as the single exponential function of Equation (7.8)

$$e(p) = \frac{E_p}{(p + \alpha)} \tag{7.25}$$

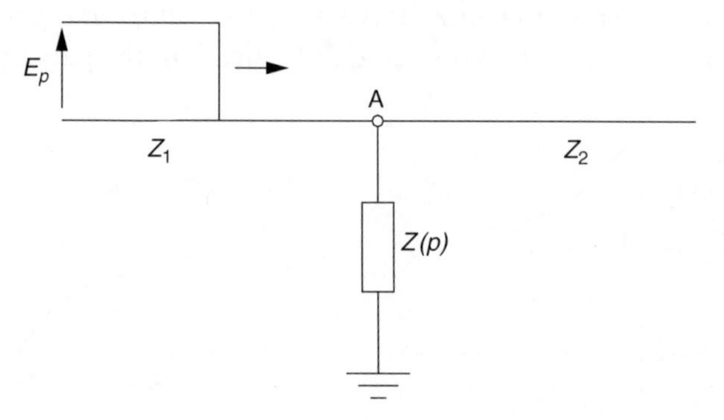


**Figure 7.6** Series connection of two transmission lines with a parallel impedance at the intersection point

and the expression for the voltage at node A can, after back transformation into the time domain, be written as

$$e_{\mathrm{A}}(p) = E_p \frac{2Z_2}{CZ_1Z_2} \frac{1}{(p+\alpha)} \frac{1}{(p+\gamma)}$$
$$= E_p \frac{2Z_2}{CZ_1Z_2} \frac{1}{\gamma-\alpha} \left( \frac{1}{p+\alpha} - \frac{1}{p+\gamma} \right) \tag{7.26}$$
$$e_{\mathrm{A}}(t) = E_p \frac{2Z_2}{Z_1 + Z_2 - \alpha(Z_1Z_2)C} \left(\mathrm{e}^{-\alpha t} - \mathrm{e}^{-\gamma t}\right) \tag{7.27}$$

with

$$\gamma = \frac{Z_1 + Z_2}{CZ_1Z_2}$$

Figure 7.7 shows the reflection and refraction of a wave travelling on a transmission line $Z_1$ toward a transmission line $Z_2$. A lumped capacitor is connected at the node connecting $Z_1$ and $Z_2$. In this example, the characteristic impedance $Z_1$ is assumed to be larger than $Z_2$. The parallel capacitance absorbs energy from the incoming wave and behaves first as a short circuit. After three times the time constant, $t > 3/\gamma$, the capacitance can be regarded as being fully charged, and therefore its influence can be neglected, as can be seen from the shape of the reflected and refracted wave. In fact, a series inductance and a parallel capacitor have the same effect on an incoming wave; they reduce the steepness of the wave front but do not reduce the amplitude of the reflected and refracted wave. The amplitude is determined by the value of the characteristic impedances $Z_1$ and $Z_2$ only. A series inductance causes a temporary rise in amplitude of the reflected wave, whereas a parallel capacitor results in a dip, but both can be used for the protection of high-voltage equipment against lightning-induced travelling waves. A parallel capacitance is often applied for the protection of transformer and generator windings. The connecting wires between the protective capacitor and the equipment to be protected must be kept as short as possible because the ever-present stray inductance, of the wires and of the loop formed by the connections, reduces the influence of the capacitance on the steepness of the wave front.

When the parallel element in Figure 7.6 is a surge-arrester, either a ZnO-type arrester or an arrester with a linear resistor in series with a spark gap, we have an element that not only *reduces* the amplitudes of the reflected and refracted waves but also *absorbs* energy from the incoming wave. The nonlinear characteristic of the metal-oxide or ZnO-arrester prevents the current flow at normal system voltage, but when the voltage rises, the

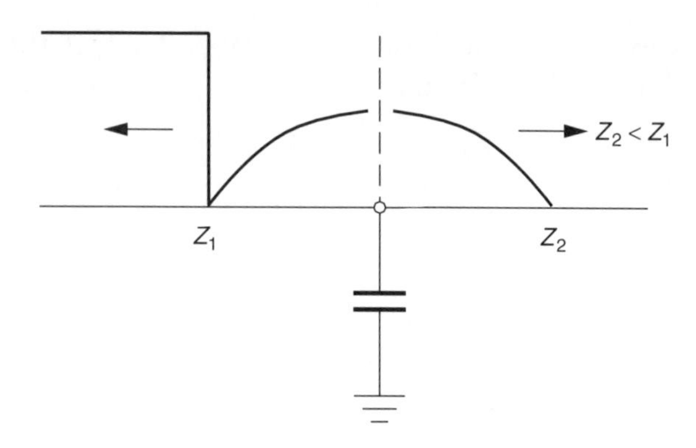


Figure 7.7 Series connection of two transmission lines with a parallel capacitor

surge arrester starts to draw current. For the classical surge arrester with spark gaps, the voltage level needs to surpass a certain voltage level until the spark gap breaks down and the resistive elements are in connection with the system. The spark gap serves, in fact, as a voltage-controlled fast switch and prevents current from flowing through the resistive elements at normal system voltage.

## *7.5 REFERENCES FOR FURTHER READING*

Greenwood, A., *Electrical Transients in Power Systems*, Wiley & Sons, New York, 1991.

Bewley, L. V., *Travelling Waves on Transmission Systems*, Chapters 2, 3, 5, Dover Publications, Mineola, New York, 1963.

Chowdhariv, P., *Electromagnetic Transients in Power Systems*, Chapter 4, Wiley & Sons, Research Studies Press Ltd, New York, 1996.

Kreuger, F. H., *Industrial High Voltage*, Chapter 7, Delft University Press, Delft, 1991.

Bergen, K. and Vogelsanger, E., "Messungen und Resultate der Blitzforschung der Jahre 1955...1963 auf dem Monte San Salvatore," *Bulletin Schweizerischer Elektrotechnischer Verein*, **56**(1), 2–22 (1965).

# 8

# Numerical Simulation of Electrical Transients

Switching actions, short-circuits, lightning strokes, and disturbances during normal operation often cause temporary overvoltages and high-frequency current oscillations. The power system must be able to withstand the overvoltages without damage to the system components. The simulation of transient voltages and currents is of great importance for the insulation coordination, correct operation, and adequate functioning of the system protection. Transient phenomena can not only occur in a time frame of microseconds or (in the case of the initial rate of rise of the transient recovery voltages and short-line faults) milliseconds (when looking at transient recovery voltages because of switching actions) but can also be present for seconds, for instance, in the case of ferro-resonance. Transients are usually composed of travelling waves on high-voltage transmission lines and underground cables, oscillations in lumped network elements, generators, transformers, and so on. To perform analytical transient calculations by hand is rather cumbersome or even impossible, and from the early days analogue scale models have been in use; the so-called *Transient Network Analyser* or TNA. The TNA consists of analogue building blocks, and transmission lines are built from lumped LC pi-sections. The first large TNA was built in the nineteen thirties and even today the TNA is used for large system studies. The availability of cheap computer power, at first mainframes, later workstations, and presently the personal computer had a great influence on the development of numerical simulation techniques. Sometimes it can be still convenient to make use of the TNA, but in the majority of the cases, computer programs, such as the widely spread and well-known *electromagnetic transients program* (EMTP) are used. These computer programs are often more accurate and cheaper than a TNA but not always easy to use!

The computer programs that were developed first were based on the techniques to compute the propagation, refraction, and reflection of travelling waves on lossless transmission lines. For each node, the reflection and refraction coefficients are computed from the values of the characteristic impedances of the connected transmission line segments. The bookkeeping of the reflected and refracted waves was done and visualised by means of a lattice diagram (see Chapter 3, *Travelling Waves*), as was first published by Bewley in 1933. Another fruitful development was the application of the Bergeron method based on the method that O. Schnyder in 1929, in Switzerland, and L. Bergeron in 1931, in France, had developed for solving pressure wave problems in hydraulic piping systems. The Bergeron method applied to electrical networks represents lumped network elements, like an $L$ or a $C$, by short transmission lines. An inductance becomes a stub lossless line with a characteristic impedance $Z = L/\tau$ and a travel time $\tau$. In the same manner, a parallel capacitance becomes a stub transmission line with a characteristic impedance $Z = \tau/C$ and a travel time $\tau$. A series capacitor, however, poses a problem.

Almost all programs for the computation of electrical transients solve the network equations in the time domain, but some programs apply the frequency domain, such as the *frequency-domain transient program* (FTP), or use the Laplace domain to solve the network equations. A great advantage of the frequency domain is that the frequency-dependent effects of high-voltage lines and underground cables are automatically included. The advantage of using the Laplace transform is that back transformation into the time domain results in a closed analytical expression. When a certain time parameter is substituted in that equation, the currents and voltages can directly be calculated for different circuit parameters. For a program with a solution directly in the time domain, the computation has to be repeated when a circuit parameter changes. The Laplace transform has, however, its drawbacks. The method cannot cope with nonlinear elements, such as surge arresters and arc models.

In this chapter, four different time domain computer programs are described:

- the Electromagnetic transient program, based on the *Nodal Analysis* from network theory;
- the MNA program, based on the *Modified Nodal Analysis* from network theory;

- the XTrans Program, based on the solution of differential and algebraic equations; and
- the MATLAB Power System Blockset.

A demo version of the XTrans Program can be downloaded from: *http://eps.et.tudelft.nl.*

## 8.1 THE ELECTROMAGNETIC TRANSIENT PROGRAM

The electromagnetic transient program (EMTP) is the creation of H. W. Dommel, who started to work on the program at the Munich Institute of Technology in the early 1960s. He continued his work at BPA (Bonneville Power Administration) in the United States. The EMTP became popular for the calculation of power system transients when Dommel and Scott-Meyer, his collaborator in those days, made the source code public domain. This became both the strength and weakness of EMTP; many people spent time on program development but their actions were not always as concerted as they should have been. This resulted in a large amount of computer code for every conceivable power system component but very often without much documentation. This problem has been overcome in the commercial version of the program, the so-called EPRI/EMTP version. Electric Power Research Institute (EPRI) has recoded, tested, and extended most parts of the program in a concerted effort and this has improved the reliability and functionality of the transient program. Circuit breaker models are an example of the extended functionality added to the program in 1987 and improved in 1997, but are not available in the still-existing public domain version of the program; the alternative transient program (ATP). Presently, the EMTP and other programs that are built on a kernel (such as electromagnetic transients for DC (EMTDC) and power system computer-aided design (PSCAD)) based on the same principles are a widely used and accepted program for the computation of electrical transients in power systems.

The EMTP is based on the application of the trapezoidal rule to convert the differential equations of the network components to algebraic equations. This approach is demonstrated in the following text for the inductance, capacitance, and lossless line.

For the inductance $L$ of a branch between the nodes $k$ and $m$, it holds

$$i_{k,m}(t) = i_{k,m}(t - \Delta t) + \frac{1}{L}\int_{t-\Delta t}^{t} (v_k - v_m)\,\mathrm{d}t \tag{8.1}$$

Integration by means of the trapezoidal rule gives the following equations

$$\begin{aligned} i_{k,m}(t) &= \frac{\Delta t}{2L}(v_k(t) - v_m(t)) + I_{k,m}(t - \Delta t) \\ I_{k,m}(t - \Delta t) &= i_{k,m}(t - \Delta t) + \frac{\Delta t}{2L}(v_k(t - \Delta t) - v_m(t - \Delta t)) \end{aligned} \tag{8.2}$$

For the capacitance C of a branch between the nodes $k$ and $m$, it holds

$$v_k(t) - v_m(t) = v_k(t - \Delta t) - v_m(t - \Delta t) + \frac{1}{C}\int_{t-\Delta t}^{t} i_{k,m}(t)\,\mathrm{d}t \tag{8.3}$$

Integration by means of the trapezoidal rule gives the following equations

$$\begin{aligned} i_{k,m}(t) &= \frac{2C}{\Delta t}(v_k(t) - v_m(t)) + I_{k,m}(t - \Delta t) \\ I_{k,m}(t - \Delta t) &= -i_{k,m}(t - \Delta t) - \frac{2C}{\Delta t}(v_k(t - \Delta t) - v_m(t - \Delta t)) \end{aligned} \tag{8.4}$$

For a single-phase lossless line between the terminals $k$ and $m$, the following equation must be true.

$$u_m(t - \tau) + Zi_{m,k}(t - \tau) = u_k(t) - Zi_{k,m}(t) \tag{8.5}$$

$\tau$ the travel time (s)
$Z$ the characteristic impedance ($\Omega$)

In words, the expression $u + Zi$ encountered by an observer when he leaves the terminal $m$ at time $t - \tau$ must still be the same when he arrives at terminal $k$ at time $t$. From Equation (8.5), the following two-port equations can be deduced.

$$\begin{aligned} i_{k,m}(t) &= \frac{u_k(t)}{Z} + I_k(t - \tau) \\ I_k(t - \tau) &= -\frac{u_m(t - \tau)}{Z} - i_{m,k}(t - \tau) \\ i_{m,k}(t) &= \frac{u_m(t)}{Z} + I_m(t - \tau) \\ I_m(t - \tau) &= -\frac{u_k(t - \tau)}{Z} - i_{k,m}(t - \tau) \end{aligned} \tag{8.6}$$

The resulting models for the inductance, capacitance, and the lossless line are shown in Figure 8.1.

They consist of current sources, which are determined by current values from previous time steps, and resistances in parallel. Thus a network can

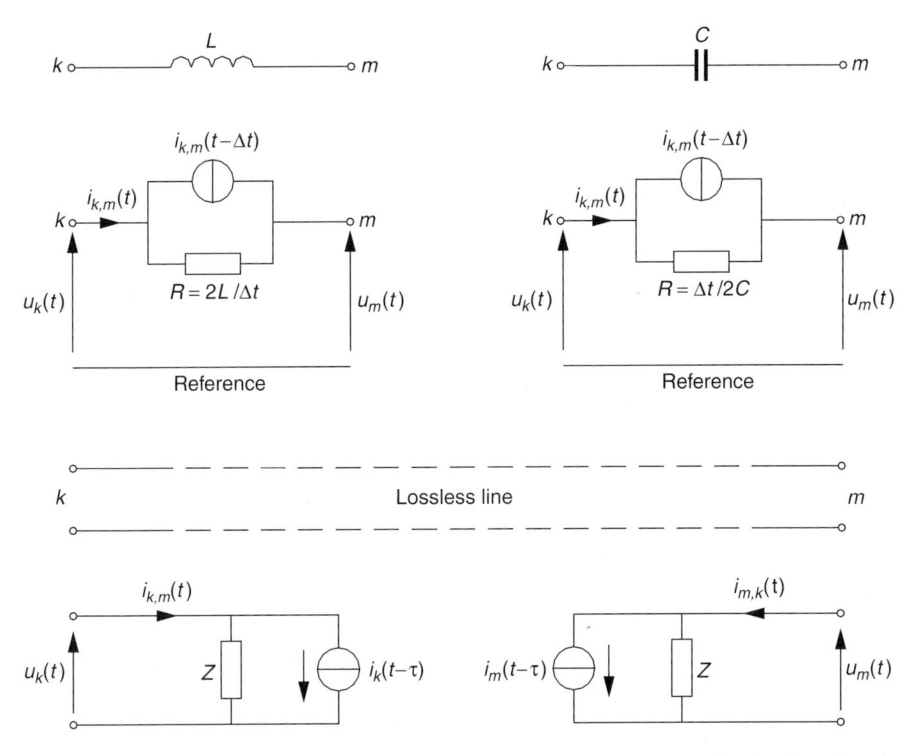


**Figure 8.1** EMTP representation of an inductance, capacitance, and a lossless line by current sources and parallel resistances. $\Delta t$ is the step size of the computation. $\tau$ is the travel time of the line. $Z$ is the characteristic impedance of the line

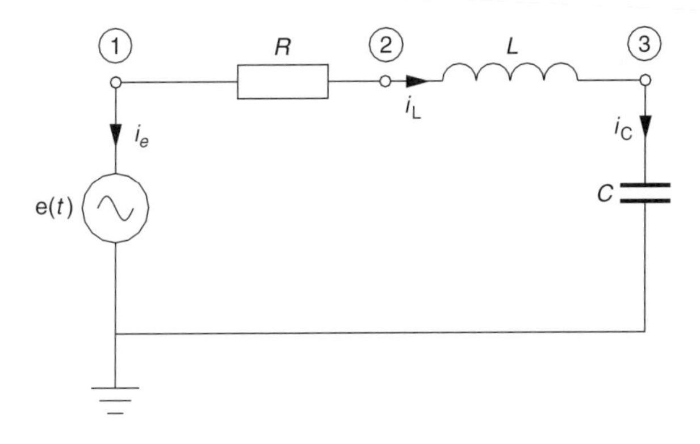


**Figure 8.2** Sample RLC circuit

be built up of current sources and resistances by using the equivalent circuits as shown in Figure 8.1. This approach will be demonstrated on the sample RLC network that is shown in Figure 8.2.

By means of the equivalent models for the inductance and the capacitance, as depicted in Figure 8.1, and the replacement of the voltage source

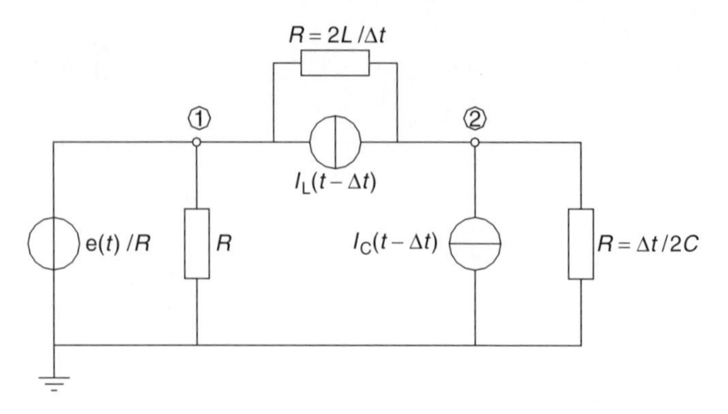


**Figure 8.3** Equivalent EMTP circuit

and the series impedance by a current source with a parallel resistance, the RLC circuit can be converted into the equivalent circuit as shown in Figure 8.3. To compute the unknown node voltages, a set of equations is formulated by using the nodal analysis (NA) method.

$$\begin{bmatrix} \frac{1}{R} + \frac{\Delta t}{2L} & -\frac{\Delta t}{2L} \\ -\frac{\Delta t}{2L} & \frac{\Delta t}{2L} + \frac{2C}{\Delta t} \end{bmatrix} \begin{bmatrix} u_1(t) \\ u_2(t) \end{bmatrix} = \begin{bmatrix} \frac{e(t)}{R} \\ 0 \end{bmatrix} - \begin{bmatrix} I_L(t-\Delta t) \\ I_C(t-\Delta t) - I_L(t-\Delta t) \end{bmatrix} \quad (8.7)$$

In general, the following equations hold

$$\boldsymbol{Yu} = \boldsymbol{i} - \boldsymbol{I} \quad (8.8)$$

$\boldsymbol{Y}$ the nodal admittance matrix
$\boldsymbol{u}$ the vector with unknown node voltages
$\boldsymbol{i}$ the vector with current sources
$\boldsymbol{I}$ the vector with current sources, that are determined by current values from previous time steps

The actual computation works as follows:

- The building up and inversion of the *Y*-matrix. This step has to be taken only once. However, when switching occurs, this step has to be repeated while the topology of the network changes;

- The time-step loop is entered and the vector of the right-hand side of Equation (8.8) is computed after a time step $\Delta t$;

- The set of linear equations is solved by means of the inverted matrix $Y$ and the vector with the nodal voltages $u$ is known; and
- The vector in the right-hand side of Equation (8.8) is computed for a time step $\Delta t$ further in time and we continue this procedure till we reach the final time.

The advantages of the 'Dommel–EMTP' method, as described in the previous section, among others are

- simplicity (the network is reduced to a number of current sources and resistances of which the $Y$-matrix is easy to construct) and
- robustness (the EMTP makes use of the trapezoidal rule, which is a numerically stable and robust integration routine).

However, the method has some disadvantages too:

- A voltage source poses a problem. This becomes clear from the sample RLC circuit; a small series resistance will result in an ill-conditioned $Y$-matrix (see Equation (8.7));
- It is difficult to change the computational step size dynamically during the calculation, the resistance values and the current sources should be recomputed at each change [see Equation (8.7)] that entails $Y$-matrix re-inversion. This is time-consuming for larger networks; and
- The $Y$-matrix is ill-conditioned. The resistances in the representations of the capacitance and the inductance (see Figure 8.1 and Equation (8.7)) are treated oppositely with regard to the computational step size $\Delta t$. Thus if the computational step size is decreased, it has a contrary effect on the inductances and capacitances. This can lead to numerical instabilities.

The arc models within the EMTP are implemented by means of the compensation method. The nonlinear elements are essentially simulated as current injections, which are superimposed on the linear network, after a solution without the nonlinear elements has been computed first.

The procedure is as follows: the nonlinear element is open-circuited and the Thevenin voltage and Thevenin impedance are computed. Now, the two following equations have to be satisfied. Firstly, the equation of the linear part of the network (the instantaneous Thevenin equivalent circuit as seen from the arc model).

$$V_{\mathrm{th}} - iR_{\mathrm{th}} = iR \tag{8.9}$$

$V_{\mathrm{th}}$ the Thevenin (open-circuit) voltage (V)
$R_{\mathrm{th}}$ the Thevenin impedance (Ω)
$i$ the arc current (A)
$R$ the arc resistance (Ω)

Secondly, the relationship of the nonlinear element itself. Application of the trapezoidal method of integration yields for the arc resistance at the simulation time $t$:

$$R(t) = R(t - \Delta t) + \frac{\Delta t}{2}\left(\left.\frac{\mathrm{d}R}{\mathrm{d}t}\right|_{t} + \left.\frac{\mathrm{d}R}{\mathrm{d}t}\right|_{t-\Delta t}\right) \tag{8.10}$$

$\Delta t$ the time step (s)

The $\mathrm{d}R/\mathrm{d}t$ is described by the differential equation of the arc model. To find a simultaneous solution of Equation (8.9) and Equation (8.10), the equations have to be solved by means of an iterative process (e.g. Newton–Raphson).

Therefore the solution process is as follows:

- The node voltages are computed without the nonlinear branch,
- Equation (8.9) and Equation (8.10) are solved iteratively, and
- The final solution is found by superimposing the response to the current injection $i$.

In the EMTP96, three arc models have been implemented; Avdonin–Schwarz, Urbanek, and Kopplin.

## 8.2 *THE MNA PROGRAM*

In the 1970s, Ho *et al.* introduced the modified nodal analysis (MNA) method to circumvent the great shortcoming of the Nodal Analysis method that provides an inefficient treatment of voltage sources. In the 1980s, a transient program called MNA was developed at KEMA's high-power laboratory by Van der Sluis and was based on this method. The MNA method uses state equations; the voltages across the capacitances and the current through the inductances, because both are continuous when the topology of the network changes through, for instance, a switching action.

The set of equations to solve the sample RLC circuit, as shown in Figure 8.2, formulated by means of the MNA method is described in the following text.

$$\begin{bmatrix} \frac{1}{R} & -\frac{1}{R} & 0 & 1 & 0 \\ -\frac{1}{R} & \frac{1}{R} & 0 & 0 & 0 \\ 0 & 0 & 0 & 0 & 1 \\ 1 & 0 & 0 & 0 & 0 \\ 0 & 0 & 1 & 0 & 0 \end{bmatrix} \begin{bmatrix} u_1(t) \\ u_2(t) \\ u_3(t) \\ i_e(t) \\ i_C(t) \end{bmatrix} = \begin{bmatrix} 0 & 0 \\ -1 & 0 \\ 1 & 0 \\ 0 & 0 \\ 0 & 1 \end{bmatrix} \begin{bmatrix} i_L(t) \\ u_C(t) \end{bmatrix} + \begin{bmatrix} 0 \\ 0 \\ 0 \\ e(t) \\ 0 \end{bmatrix} \tag{8.11}$$

The vector with the state variables, containing the currents through the inductances and the voltages across the capacitance, can be obtained by integration of the following equation.

$$\begin{bmatrix} i'_L(t) \\ u'_C(t) \end{bmatrix} = \begin{bmatrix} 0 & \frac{1}{L} & -\frac{1}{L} & 0 & 0 \\ 0 & 0 & 0 & 0 & \frac{1}{C} \end{bmatrix} \begin{bmatrix} u_1(t) \\ u_2(t) \\ u_3(t) \\ i_e(t) \\ i_C(t) \end{bmatrix} \tag{8.12}$$

In general, the following equations hold

$$\boldsymbol{x}'_1(t) = \boldsymbol{A}\boldsymbol{x}_2(t) \tag{8.13}$$

$$\boldsymbol{B}(t)\boldsymbol{x}_2(t) = \boldsymbol{C}\boldsymbol{x}_1(t) + \boldsymbol{f}(t) \tag{8.14}$$

$\boldsymbol{x}_1$ the $n$-vector of differential variables
$\boldsymbol{x}_2$ the $m$-vector of algebraic variables
$\boldsymbol{A}$ the $n \times m$-matrix that represents the linear relation between $x'_1$ and $x_2$
$\boldsymbol{B}(t)$ the $m \times m$-MNA matrix; the matrix can contain time-dependent elements
$\boldsymbol{C}$ the $m \times n$-matrix that represents the linear relation between $x_1$ and $x_2$
$\boldsymbol{f}(t)$ the $m$-vector with the (time-dependent) contributions from voltage and current sources

Normally, the MNA matrix will not be time-dependent. However, it will be time-dependent in the case an arc model is incorporated as a nonlinear conductance. Such a time-dependent MNA matrix should be inverted or factorised at each time step, but this is not efficient because of the very long computation times. To solve Equation (8.13) and Equation (8.14), the following method is used.

- The right-hand side vector of Equation (8.14) is calculated with the initial values of $\boldsymbol{x}_1$.
- If the MNA matrix is time-dependent, the matrix is updated and inverted or factorised.
- From Equation (8.14), $\boldsymbol{x}_2$ is computed.
- The values of $\boldsymbol{x}_1$ at the new time step are computed by solving Equation (8.13) with a numerical integration method. This process is repeated until the final simulation time is reached.

The advantages of the MNA method, as described in the earlier section, among others are as follows:

- Voltage sources are easily included (as shown in the case of the sample RLC circuit), and
- the computational step size is not present within the MNA matrix (see Equation (8.11)). If the step size is adjusted during the calculation, the MNA matrix does not need to be inverted or factorised once more (assuming that the MNA matrix is not time-dependent).

The method has one disadvantage:

- The matrices $\boldsymbol{A}$, $\boldsymbol{B}$, and $\boldsymbol{C}$ require linearity and it is necessary that the two sets of unknowns can be defined and solved separately from the corresponding set of equations. Therefore the MNA method deals with linear models only. Nonlinear diodes, arc models, and so on, can be incorporated by means of tricks only.

An arc model can be implemented within the MNA method in various ways.

The arc model can be modelled as a nonlinear conductance. However, the MNA matrix becomes time-dependent and should be inverted or factorised at each time step. This involves long computation times and is, therefore, not efficient. An alternative can be found by applying the so-called partial matrix (factor) updating. Now the changes are not made within the MNA matrix itself, but in the inverted MNA matrix or the MNA matrix factors. This introduces additional computation time for the updating, but the MNA matrix has to be inverted or factorised only once.

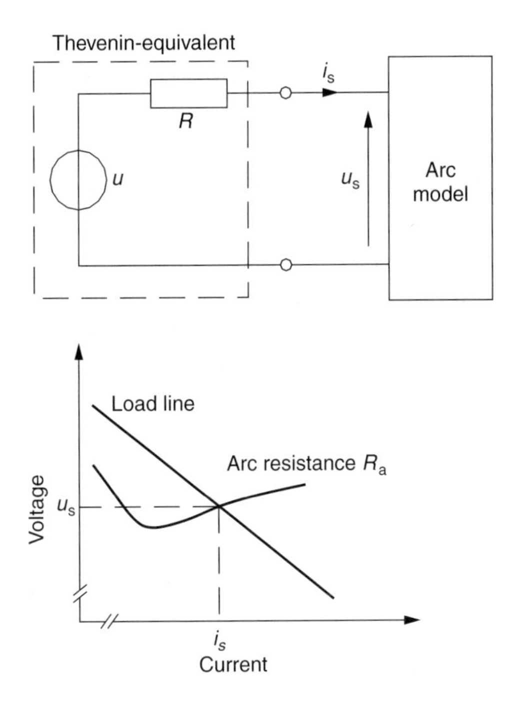


**Figure 8.4** Arc model connected to a linear network

Another approach is the one that Van der Sluis followed to implement the Mayr arc model in the MNA computer program. The arc model is treated as a voltage source. This technique is depicted in Figure 8.4. The linear network to which the arc model is connected can be replaced by its Thevenin equivalent with its corresponding load line.

Each time step, the arc resistance $R_a$ is computed from the arc models' differential equation. At the point of intersection of the momentary arc resistance and the load line, both the network and the arc model have the same voltage $u_s$ and current $i_s$. If two points of the load line, $(u_1, i_1)$ and $(u_2, i_2)$, are known, the voltage $u_s$ and current $i_s$ are computed from the following equation.

$$u_s = u_1 + \frac{(R_a i_1 - u_1)(u_2 - u_1)}{u_2 - u_1 + R_a(i_1 - i_2)} \qquad i_s = \frac{u_s}{R_a} \tag{8.15}$$

## 8.3 THE XTRANS PROGRAM

At the Power Systems Laboratory of the Delft University of Technology, Bijl and Van der Sluis developed the transient program XTrans. A demo version of the program can be downloaded from *http://eps.et.tudelft.nl*. The program was developed especially for analysis of arc–circuit interaction, which involves nonlinear elements in relation to stiff differential equations. The calculations are performed with a variable step size and adjustable accuracy of the computed currents, voltages, and conductances.

The application of the NA and MNA method has some drawbacks, especially when nonlinear elements, such as arc models are used, as described in the previous sections. A more general representation of a set of first-order differential equations and algebraic equations (DAEs) is given by:

$$F(x'(t), x(t), t) = 0 \tag{8.16}$$

The set of equations to solve the sample RLC circuit as shown in Figure 8.2 formulated by means of the DAEs is described in the following text.

$$\begin{bmatrix} i_{\mathrm{e}}(t) & + & \dfrac{u_1(t) - u_2(t)}{R} \\ \dfrac{u_1(t) - u_2(t)}{R} & - & i_{\mathrm{L}} \\ i_{\mathrm{L}} & - & i_{\mathrm{C}} \\ u_1(t) & - & \mathrm{e}(t) \\ i_{\mathrm{L}}(t)' & - & \dfrac{u_2(t) - u_3(t)}{L} \\ u_3(t)' & - & \dfrac{i_{\mathrm{C}}(t)}{C} \end{bmatrix} = 0 \tag{8.17}$$

When using the DAEs, an arc model, such as the Mayr arc model, can be implemented easily by extending the system of equations as follows:

$$\begin{pmatrix} \vdots \\ i - gu \\ \vdots \\ g' - \dfrac{1}{\tau}\left(\dfrac{u^2 g^2}{P} - g\right) \end{pmatrix} = 0 \tag{8.18}$$

The solution of the DAEs [Equation (8.16)] is achieved by using the BDF method (backward differentiation formulas method). This method was first proposed by Gear in 1971 and has since then been studied and developed by mathematicians. The solution is performed in two steps:

- Suppose the solution $x^n$ at $t = t^n$ is given, then an appropriate step size $h^n = t^{n+1} - t^n$ is selected to be able to compute an approximation of $x'(t^n+1)$, using the multistep BDF formula:

$$x^{n+1'} = \frac{1}{h^n} \sum_{i=0}^{k} \alpha_i x^{n+1-i} \tag{8.19}$$

$k$ the order of the method
$\alpha_i$ the coefficients of the method
$h$ the step size of the computation

This is called a *multistep method* because the calculation results of the last $k$ time steps are used. In the special case that $k = 1$, Equation (8.19) yields the one-step Euler formula:

$$x^{n+1'} = \frac{x^{n+1} - x^n}{h^n} \tag{8.20}$$

The determination of the step size and the coefficients $\alpha_i$ is based on strategies that are described in detail in the book *Numerical solutions of Initial-Value Problems in Differential Algebraic Equations* by Brennan, Campbell, and Petzold.

- In Equation (8.16), $x'$ is replaced by the backward difference [Equation (8.19)] and this gives us the following nonlinear set of equations.

$$F(x^{n+1'}, x^{n+1}, t^{n+1}) = g(x^{n+1}, t^{n+1}) = 0 \tag{8.21}$$

These equations can be solved by the Newton-Raphson method that consists of the following iteration formulas:

$$J \Delta x_i^{n+1} = g(x_i^{n+1}, t^{n+1}) \tag{8.22}$$

$$x_{i+1}^{n+1} = x_i^{n+1} + \Delta x_i^{n+1} \tag{8.23}$$

$J$ the Jacobian matrix of $g$
$x_i^{n+1}$ the $i$th approximation of $x^{n+1}$

This iteration process is started with initial guesses $x_0^{n+1}$ to make sure that the desired solution is reached. These initial guesses are found by extrapolating the earlier solution points. Subsequently the mismatch vector [i.e. the right-hand side of Equation (8.22)] is computed by filling in these values in $g$. The correction vector, $\Delta x_i^{n+1}$, is solved from Equation (8.22). The vector $x^{n+1}$ is updated as shown in Equation (8.23) and the process is repeated until convergence has been achieved. This solution process may seem time-consuming compared with the NA or MNA method but this is not necessarily the case. The use of a multistep method makes it possible to take larger time steps than can be done with a single-step method. The possibility to adjust the step size during calculation makes it possible to solve stiff equations and avoid the cumulation of errors.

To solve Equation (8.16), a consistent set of initial values should be available. Note that these initial values are not only needed at the start of the calculation but also after switching operations; voltages and currents may not be constant across the switching boundaries. Letting the user specify these initial values is normally impracticable, especially after switching operations. This means that the XTrans program has to calculate these initial values. Because of the nonlinearity of the set of equations involved, an iterative method such as the Newton–Raphson method should be applied. The drawback of this and other iterative methods is that they will converge to the proper solution only if a reasonable guess of that solution can be given. The problem thus remaining is to specify this initial guess. For this purpose, the set of equations is linearised and the solution is found with Opal's two-step integration procedure based on numerical Laplace inversion. (See the references for further reading at the end of this chapter).

To illustrate this, the Laplace transformation will be applied on a case with inconsistent initial values. In Figure 8.5, a sample network is depicted with inconsistent initial conditions when the (ideal) switch closes. For the network, the following settings are used: $C_1 = 0.145\ \mu F$, $C_2 = 0.146\ \mu F$, $L = 14.75$ mH, $R = 25\ \Omega$, and $C_1$ is charged at 100 V. When the switch closes, the inductance can be regarded as an open branch (conservation of flux).

The Laplace transformation of the voltage $V$ after switching can be written as:

$$V(s) = \frac{C_1 V_{C1}(0^-) + C_2 V_{C2}(0^-)}{pC_1 + pC_2} = \frac{C_1 V_{C1}(0^-)}{p(C_1 + C_2)} \tag{8.24}$$

Back-transformation to the time domain leads to the following expression for the voltage:

$$v(t) = \frac{C_1 V_{C1}(0^-)}{C_1 + C_2} = 49.8\ \text{V} \tag{8.25}$$

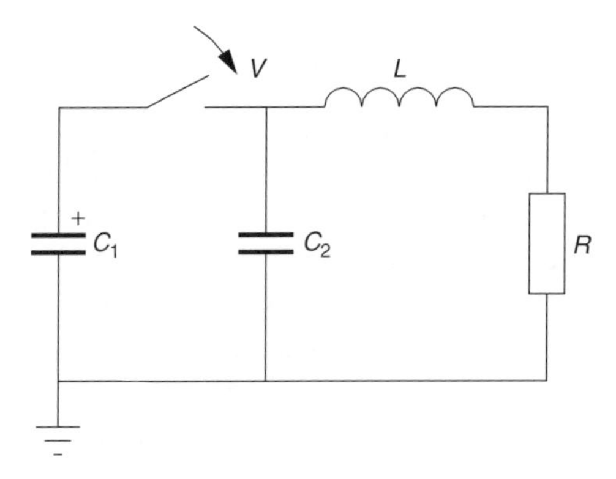


Figure 8.5 Sample network with inconsistent initial conditions

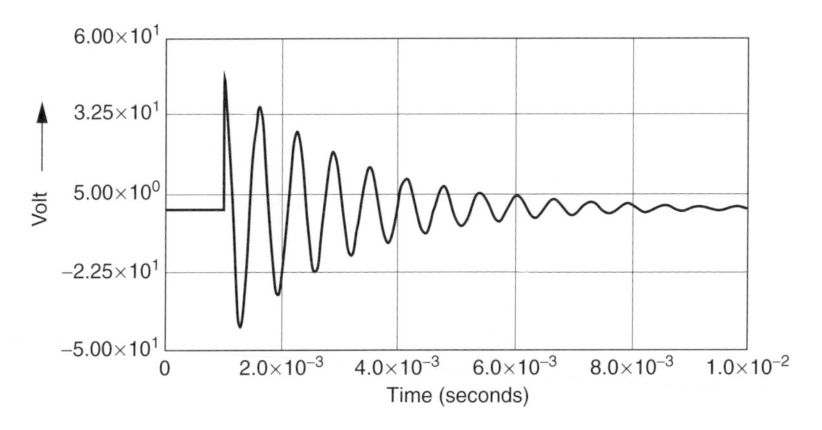


Figure 8.6 Voltage on the capacitor $C_2$

The response of the circuit, as computed by XTrans, is shown in Figure 8.6; the instantaneous voltage jump of 49.8 V can be clearly recognised. This example shows that the inverse Laplace transformation can be used successfully to solve a system with inconsistent initial conditions, which can lead to Dirac pulses (as in this example)

An electrical network consists of nodes connected to each other by elements. Elements and nodes both have their own related equations and unknowns. The equations related to the nodes represent the connections between the elements. They are obtained by applying Kirchhoff's current law (KCL) to every node except the datum or reference node. The corresponding unknowns are the voltages at these nodes. The NA method uses these equations and unknowns only, but the MNA method gives network elements the possibility to introduce extra unknowns and equations. These equations have to be linear and algebraic when the MNA method is used, but can also be nonlinear and first-order differential if the BDF method is used.

As demonstrated, the solution of Equation (8.16) is equal to the solution of Equation (8.21) when applying the iteration formulas (8.22) and (8.23). This iteration process communicates with the 'outside' world by making a guess for $x^{n+1}$ and expecting a new mismatch vector in return. With the guesses for the terminal voltages and the introduced unknowns of an element, the mismatches for the introduced equations and the currents at the terminals are calculated. These currents are used in the main program to calculate the mismatches corresponding to Kirchhoff's equations.

The requirements of an element model to work with the BDF solution method can be summarized as follows:

- The additional equations introduced by an element model must be algebraic equations or first-order differential equations. These additional equations must add new information to the set of equations and

- The element models must be able to return guesses for the currents at their terminals. The data available for this purpose are the guesses of the introduced unknowns, the guesses of the node voltages at the terminals, and the first-order derivatives of these unknowns (obtained with Equation (8.19)).

In Table 8.1, it is shown how unknowns and equations are introduced, and how terminal currents are calculated and returned for some basic elements.

XTrans is a computer program that uses libraries that contain information about the behaviour of element models, as shown in Figure 8.7. The computer program contains the general behaviour of the element models, and the libraries fill in the specific properties. This is realized with the use of OOP (object-oriented programming). The basic properties of OOP are outlined first.

**Table 8.1** Unknowns, equations, and terminal currents for basic elements between node a and node b

| Element | Unknown | New equation | Terminal current |
|---|---|---|---|
| Resistance | | | $\pm(U_a - U_b)/R$ |
| Inductance | $i_L$ | $i_L' = (U_a - U_b)/L$ | $\pm i_L$ |
| Capacitance | $i_C$ | $U_a' - U_b' = i_C/C$ | $\pm i_C$ |
| Voltage source | $i_E$ | $U_a - U_b = E$ | $\pm i_E$ |
| Current source | | | $\pm I$ |

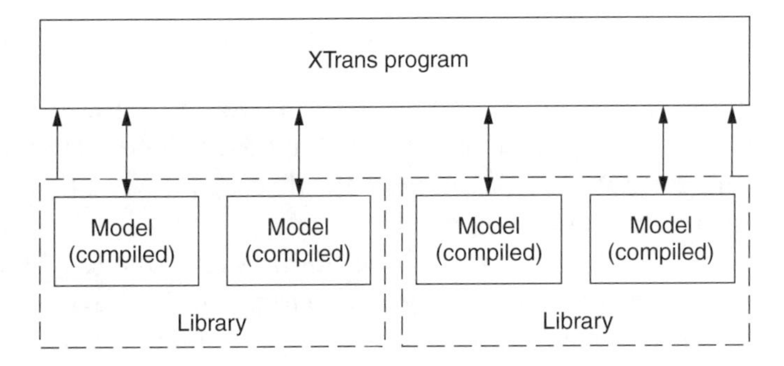


**Figure 8.7** Structure of the XTrans program

In computer programming, records are used to collect data, which are closely related to each other, in one new variable. The idea of OOP is to extend this related data by the addition of related methods. (A function or procedure, operating on that specific data is indicated by a method.) Such a group of related data and methods is called an object class. Each object class can be a base class for one or more derived classes. A derived class inherits all the functionality of its base class and adds new, more specific, functionality to it. A derived class can also overload the virtual functions of its base class. This property allows a derived class to change the contents of a function that was already defined in its base class.

The common behaviour of element models that must be known by the main program is defined in a general base class 'element.' This base class defines virtual functions for all different instants in the process of defining a network, calculating the results, and visualising these results. There is, for instance, a function that returns the image of the element for drawing purposes. There are functions for the various steps in obtaining the initial values, for calculating the mismatches and the terminal currents, functions for returning output data, and so on. The element models can now be defined by deriving new classes from this base class and overloading those virtual functions that are of interest for these elements. However, not every virtual function needs to be overloaded. These new classes can then be compiled and placed in a library.

The advantage of this approach is the possibility of having intense interaction with the coupled element models by defining many virtual functions in the base class. A further advantage is that the source code for new element models is compiled instead of interpreted, which results in faster execution.

## 8.4 THE MATLAB POWER SYSTEM BLOCKSET

After the introduction of the power system blockset (PSB), for modelling and simulating electric power systems within the MATLAB Simulink environment, the general-purpose mathematical program MATLAB is a suitable aid for the simulation of power system transients. The Power System Blockset is developed at TEQSIM Inc. and Hydro-Québec. Simulink is a software package for modelling, simulating, and analysing dynamic systems. It provides a graphical user interface for building models as block diagrams. The PSB block library contains Simulink blocks that represent common components and devices found in electrical power networks. The measurement blocks and the controlled sources in the PSB block library act as links between electrical signals (voltages across elements and currents flowing through lines) and Simulink blocks (transfer functions) and vice versa respectively.

Simulink is based on the interconnection of blocks to build up a system. Each block has three general elements: a vector of inputs $u$, a vector of outputs $y$, and a vector of state variables $x$ (Figure 8.8).

Before the computation, the system is initialised; the blocks are sorted in the order in which they need to be updated. Then by means of numerical integration with one of the implemented ODE (ordinary differential equation) solvers, the system is simulated. The computation consists of the following steps:

- The block output is computed in the correct order;
- The block calculates the derivatives of its states based on the current time, the inputs, and the states; and
- The derivative vector is used by the solver to compute a new state vector for the next time step. These steps continue until the final simulation time is reached.

The PSB block library contains Simulink blocks that represent common components and devices found in electrical power networks. Therefore systems can be built up, consisting of both Simulink and PSB blocks. However, the measurement blocks and the controlled sources in the

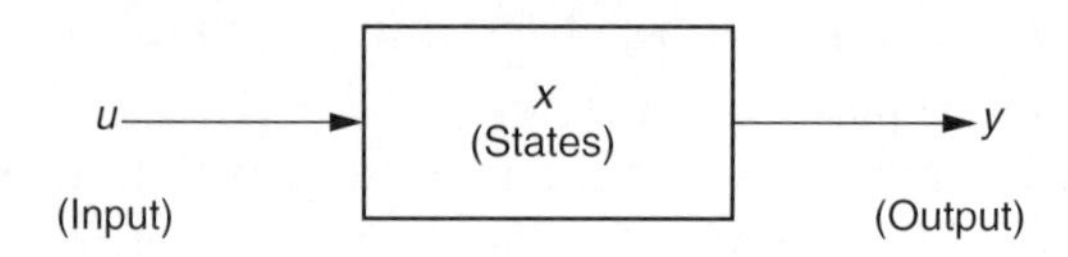


**Figure 8.8** Simulink Block

PSB block library act as links between electrical signals (voltages across elements and currents flowing through lines) and Simulink blocks (transfer functions) and vice versa respectively. Before the computation, the system is initialised; the state-space model of the electric circuit is computed, and the equivalent Simulink system is built up. The computation itself is analogous to the previously mentioned Simulink computational process.

The sample RLC circuit, previously used in this chapter, built up in the PSB is shown in Figure 8.9.

It is important to note that the arrows in the diagram do not indicate causality, as is the case in the Simulink block diagrams. The state space model of this sample circuit is described in the following text.

$$\underline{\dot{x}} = \begin{bmatrix} -\frac{R}{L} & -\frac{1}{L} \\ \frac{1}{C} & 0 \end{bmatrix} \underline{x} + \begin{bmatrix} \frac{1}{L} \\ 0 \end{bmatrix} V_{\mathrm{AC}}$$

with

$$\underline{x} = \begin{bmatrix} i_{\mathrm{L}} \\ u_{\mathrm{C}} \end{bmatrix} \tag{8.26}$$

MATLAB contains a large number of ODE solvers, some with fixed and others with variable step size. The ode15s solver, which can be used to solve stiff systems, is a variable order solver based on the numerical differentiation formulas (NDF's). These are more efficient than the BDF's, though a couple of the higher-order formulas are somewhat less stable.

The arc models can be modelled as voltage-controlled current sources. This approach is visualised in Figure 8.10, where both the Mayr arc model block and the underlying system are shown as implemented in the arc model blockset, which is freely available on *http://eps.et.tudelft.nl*.

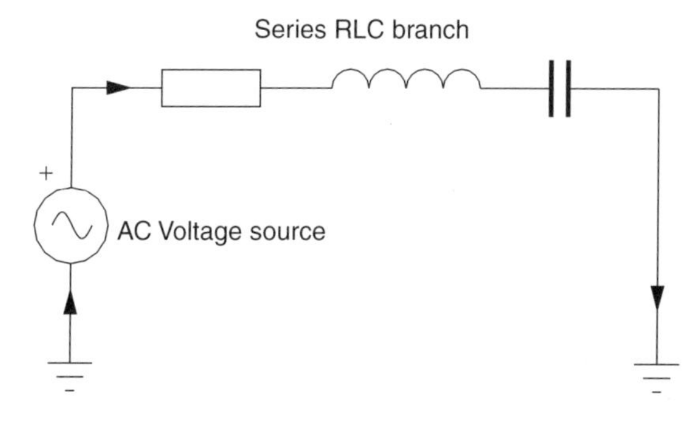


**Figure 8.9** Sample RLC circuit in the MATLAB PSB

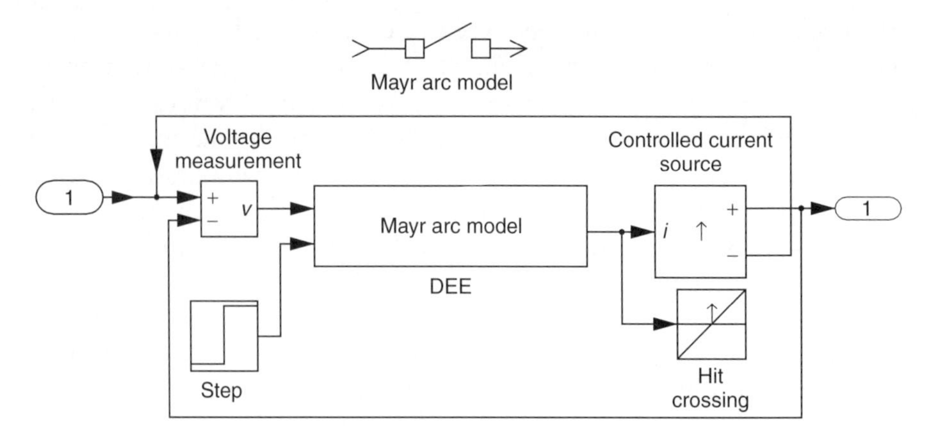


**Figure 8.10** Implementation of the Mayr arc model in Matlab Simulink/Power System Blockset

The equations of the Mayr arc model have been incorporated by means of the Simulink DEE (differential equation editor) block, as shown in Figure 8.11.

Therefore, the following system of equations is solved:

$$\left.\begin{array}{c}\dfrac{\mathrm{d}x(1)}{\mathrm{d}t} = \dfrac{u(2)}{\tau}\left(\dfrac{\mathrm{e}^{x(1)}u(1)^2}{P} - 1\right) \\ y = \mathrm{e}^{x(1)}u(1)\end{array}\right| \begin{array}{c}\dfrac{\mathrm{d}\ln g}{\mathrm{d}t} = \dfrac{u(2)}{\tau}\left(\dfrac{gu^2}{P} - 1\right) \\ i = gu\end{array} \quad (8.27)$$

- $x(1)$ the state variable of the differential equation, which is the natural logarithm of the conductance: $\ln(g)$.
- $x(0)$ the initial value of the state variable, that is, the initial value of the arc conductance: $g(0)$.
- $u(1)$ the first input of the DEE block which is the arc voltage: $u$.
- $u(2)$ the second input of the DEE block, which represents the contact separation of the circuit breaker: $u(2) = 0$ when the contacts are closed, and $u(2) = 1$ when the contacts are being opened.
- $y$ the output of the DEE block, which is the arc current: $i$.
- $g$ the arc conductance
- $u$ the arc voltage
- $i$ the arc current
- $\tau$ the arc time constant
- $P$ the cooling power

$\tau$ and P are the free parameters of the Mayr arc model, which can be set by means of the dialogue, as depicted in Figure 8.12, that appears when the Mayr arc model block is double clicked.

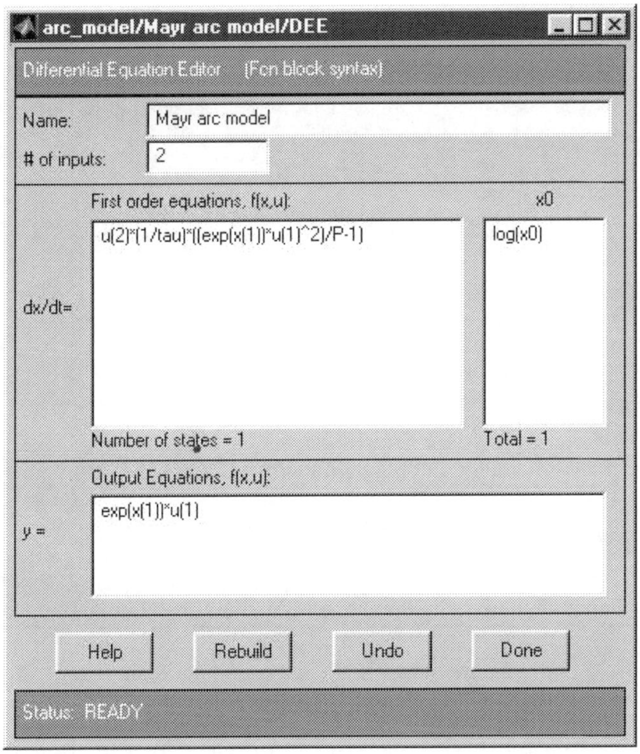

**Figure 8.11** Mayr equation in the Simulink Differential Equation editor

The Simulink 'hit crossing' block detects when the input, in this case the current, crosses the zero value. By adjusting the step size, this block ensures that the simulation detects the zero crossing point. This is important because the voltage and current zero crossing of the arc, which behaves as a nonlinear resistance, must be computed accurately.

The Simulink 'step' block is used to control the contact separation of the circuit breaker. A step is made from a value zero to one at the specified contact separation time. When the contacts are still closed, the following differential equation is solved:

$$\frac{\mathrm{d}\ln g}{\mathrm{d}t} = 0 \tag{8.28}$$

Therefore, the arc model behaves as a conductance with the value $g(0)$. From the contact separation time on, the Mayr equation is being solved:

$$\frac{\mathrm{d}\ln g}{\mathrm{d}t} = \frac{1}{\tau}\left(\frac{gu^2}{P} - 1\right) \tag{8.29}$$

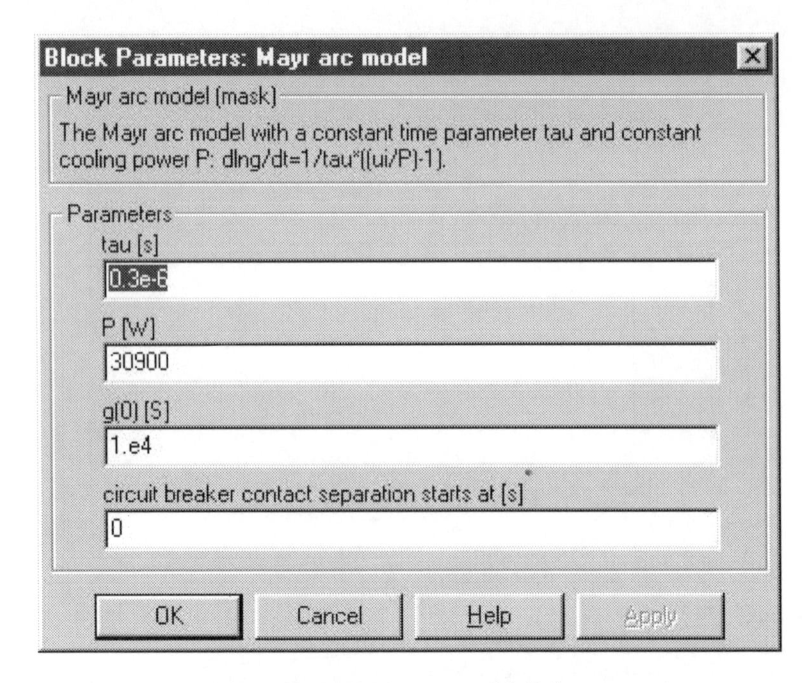


**Figure 8.12** Mayr arc model dialogue

Both the initial value of the arc conductance $g(0)$ and the time at which the contact separation of the circuit breaker starts are specified by means of the arc model dialogue, as displayed in Figure 8.12.

## *8.5 REFERENCES FOR FURTHER READING*

Barthold, L. O. and Carter, G. K., "Digital travelling wave solutions," *AIEE T-PAS*, **80**, 812–820 (1961).

Bijl, N. and van der Sluis, L., "New approach to the calculation of electrical transients," *European Transactions on Electrical Power (ETEP)*, Vol. **8**, No. 3, Part I: Theory, Part II: Applications, Eurel Publication, VDE Verlag, Berlin, 1998, 175–179, 181–186.

Brennan, K. E., Campbell, S. L., and Petzold, L. R., *Numerical Solutions of Initial-value Problems in Differential-Algebraic Equations*, North Holland, New York, 1989.

Chung-Wen, Ho, Ruehlti, A. E., and Brennan, P. A., "The modified nodal approach to network analysis," *IEEE T-CAS*, **CAS-22**(6), 504–509 (1975).

Dommel, H. W., "A method for solving transient phenomena in multiphase systems," *Proceedings of the* 2nd *PSCC*, Stockholm, Report 5.8, 1966.

Dommel, H. W., "Digital computer solution electromagnetic transients in single and multiphase networks," *IEEE Transactions on Power Apparatus and Systems*, **PAS-88**(4), 388–399 (1969).

Dommel, H. W., "Nonlinear and time-varying elements in digital simulation of electromagnetic transients," *IEEE T-PAS*, 2561–2567 (1971).

Gear, C. W., "Simultaneous numerical solutions of differential-algebraic equations," *IEEE Transactions on Circuit Theory*, **18**(1), 89–95 (1971)

Opal, A. and Vlach, J., Consistent initial conditions of linear switched networks, *IEEE Transactions on Circuits and Systems*, **CAS-37**(3), 364–372 (1990).

Phadke, A. G., Scott, Meyer W., and Dommel, H. W., "Digital simulation of electrical transient phenomena," *IEEE tutorial course*, EHO 173-5-PWR, IEEE Service Center, Piscataway, New York, 1981.

Phaniraj, V. and Phadke, A. G., "Modelling of circuit breakers in the electromagnetic transients program," *Proceedings of PICA*, 476–482 (1987).

The Mathworks, Power System Blockset, for use with Simulink User's Guide, Version 1, The Mathworks, Natick, Massachusetts, 1999.

The Mathworks, Simulink, Dynamic System Simulation for MATLAB – Using Simulink, Version 3 The Mathworks, Natick, Massachusetts, 1999.

Singhal, K. and Vlach, J., "Computations of time domain response by numerical inversion of the Laplace transform," *Journal of the Franklin Institute*, **299**(2), 109–126 (1975).

Sybille, G. *et al.*, " Theory and applications of power system blockset: a MATLAB/Simulink-based simulation tool for power systems," *IEEE Winter Meeting*, Singapore, January, 2000.

Van der Sluis, L., Rutgers, W. R., and Koreman, C. G. A., "A physical arc model for the simulation of current zero behaviour of high-voltage circuit breakers," *IEEE Transactions on Power Delivery*, 7(2), 1016–1022 (1992).

Van der Sluis, L., "A network independent computer program for the computation of electrical transients," *IEEE T-PRWD*, **2**, 779–784 (1987).

# 9

# Insulation Coordination, Standardisation Bodies, and Standards

Each component of the power system, whether it is a simple string insulator or a large power transformer, is continuously stressed by the system operating voltage at power frequency. Occasionally overvoltages occur, having a peak value exceeding the peak value of the system operating voltage. The overvoltages can be divided into three categories:

- Overvoltages caused by lightning discharges (see Chapter 7, *Lightning-Induced Transients*),
- Switching overvoltages (see Chapter 5, *Switching Transients*), and
- Sustained AC overvoltages.

The lightning discharge is a current injection either in the vicinity of a line or substation, in the transmission line tower, or directly on the line. The voltages developed across the power system components depend on the characteristic impedance of the components. The waveforms of the lightning-induced overvoltages are all different in amplitude and capricious in shape. For the sake of testing the dielectric impulse strength of power apparatus in the high-voltage laboratory, the impulse voltage wave has been standardised with the 1.2/50-μs waveform.

Switching overvoltages result from switching operations in the network. Switching of a short-circuit current, clearing a short-line fault, disconnecting unloaded transformers, disconnecting or connecting unloaded distribution cables or transmission lines, all result in damped oscillatory voltages, the so-called transient recovery voltage or TRV.

AC overvoltages occur when the 50- or 60-Hz operating voltage temporarily attains a high value at the receiving end of a transmission line after a sudden loss of load. The resistive and reactive voltage drop disappears and the overvoltage stresses the system until the operating voltage is restored. Another situation for AC overvoltages to occur is in the case of a single-phase-to-ground fault in an isolated neutral system. The healthy phases rise from phase voltage until the $\sqrt{3}$ times higher line voltage. Also the capacitance of an unloaded distribution cable in combination with the inductance of a power transformer or generator can increase the system voltage resulting in a sustained AC overvoltage.

The IEC defines insulation coordination as 'the selection of the dielectric strength of equipment in relation to the voltages that can appear on the system for which the equipment is intended and taking into account the service environment and the characteristics of the available protective devices.' This means that the insulation level of the power system components should be such that its tolerance level under transient voltages is higher than the level to which transient voltages will be limited by protective devices. Power system components have to withstand the influence of overvoltages for a lifetime of thirty to fifty years. For this reason, tests have been specified and standards have been written. Equipment is designed according to the specification of the standards, but it remains often, if not always, necessary to demonstrate the adequacy of the design of such equipment by testing in the high-voltage and in the high-power laboratory. The testing laboratory issues a *type test certificate*, a document indicating that adequate confidence is provided and that a duly identified product design complies with the requirements of a specific standard. A *certificate* contains a record of a series of type tests carried out strictly in accordance with a recognised standard and the equipment tested has fulfilled the requirements of this standard. The relevant ratings assigned by the manufacturer are endorsed by the testing laboratory. The certificate is applicable only to the equipment tested and contains the essential drawings and a description of the equipment tested.

Standards for the testing of power system components are issued by IEC and ANSI/IEEE. They are based on extensive system studies carried out by the numerous CIGRE working groups of the different study committees. The independent high-power laboratories are united in the short-circuit testing liaison (STL). STL issues guides for short-circuit testing of high-voltage apparatus.

## *9.1 THE INTERNATIONAL ELECTROTECHNICAL COMMISSION – IEC*

The revolution in electrical engineering took place at the turn of the nineteenth century. In a rather short period of time the transformer was invented, electric motors and generators were designed, and the step from DC to AC transmission was made. Lighting, at first but rapidly, the versatile application of electrical power completely changed society. In this early period, independent operating power companies used different voltage levels and operated their system at various frequencies. Electrical engineers were among the first to realize that international standardisation would become necessary in the modern world and a number of congresses were held on the topic at the end of the nineteenth and beginning of the twentieth century. It was agreed that a permanent organisation capable of carrying out international standardisation in a methodological and continuous manner was necessary.

The International Electrotechnical Commission was founded in 1906 because of the resolution passed by the Chamber of Government delegates at the International Electrotechnical Congress held in St. Louis (USA) in 1904. The first president was Lord Kelvin and the first general secretary was Charles le Maistre. At the time of the formation of the organisation, a Central Office was set up in London, which was to remain the seat of the IEC until the Central Office was transferred to Geneva in 1947. In that same year the IEC became affiliated with the International Organisation for Standardisation (ISO) as its electrical division. General meetings have been held yearly since 1947.

The membership consists of more than fifty participating countries. Regular members are the National Committees of the participating countries. This enables the widest degree of consensus on standardisation work to be reached at an international level. It is up to the National Committees to align their policies accordingly at the international level.

The IEC's principal activity is developing and publishing international standards and technical reports. The international standards serve as a basis for national standardisation and as a reference when drafting international tenders and contracts. A component or system manufactured to IEC standards and manufactured in one country can be sold and used in other countries.

The preparation of a new IEC standard or the revision of an existing standard takes place in different stages. In the *preliminary stage*, data is collected or round-robin tests are made. In the *proposal stage*, a proposal is generally made by industry via a national committee and is communicated to the members of the appropriate technical committee (TC) or study committee (SC). If a majority vote is positive, five members undertake to participate actively in the work and nominate experts. Then in the *preparatory stage*, a working draft (WD) is prepared by a working group. In the *committee stage* this document is submitted to the national committees for comment and ballot if the document is intended to be published as a technical report. After the *enquiry* and the *approval stage*, during which the principal members have to approve the document with a majority of two thirds of the votes, the final document is published by the Central Office in Geneva. In 1997, the IEC introduced a new numbering system for all its international standards, guides, and technical reports. A block of numbers ranging from 60 000 to 79 999 is now being used. Existing publications also adopt this new numbering system. For example, IEC 427 is now referred to as IEC 60427.

## 9.2 THE AMERICAN NATIONAL STANDARDS INSTITUTE–ANSI

With the application of the first circuit breakers in the early power systems, the development of standards for rating, testing, and manufacturing of high-voltage circuit breakers began. The initiative was taken by a number of engineering and manufacturers trade organisations, such as AIEE (American Institute of Electrical Engineers), NELA (National Electric Light Association), the Electric Power Club, which later became NEMA (National Electrical Manufacturers Association), AEIC (Association of Edison Illuminating Companies), and EEI (Edison Electric Institute). Until 1940, these organisations published several standardisation proposals mainly concerning rating and testing. In 1945, this series was issued as an approved American Standard with the familiar C37 number identification. The AIEE Switchgear Committee initiated the development of a circuit breaker rating method based on symmetrical interrupting currents in 1951.

## 9.3 THE CONFÉRENCE INTERNATIONALE DES GRANDS RÉSEAUX ÉLECTRIQUES À HAUTE TENSION–CIGRÉ

In the beginning of the twentieth century, the transmission voltages grew rapidly to reduce transmission losses. To improve operating efficiency, power systems began to interconnect. The reserve power or spinning reserve could be shared and capital expenditure could be reduced. One utility could supply power to another utility whose load was high at times when its own load was low. One of the prime technical problems in the past was the parallel operation of electrical machines. CIGRÉ was founded in 1921 and in 1927 the organisation was structured by means of study committees (SC). In the past, the highest transmission voltage level of 150 kV was in transition to 220 kV. CIGRÉ took actively part in the technical journey of rapidly expanding power systems. Especially after the Second World War the economies were growing rapidly and the utilities had to meet a constantly increasing load. An important aspect of those times was the technological and engineering issues of power system operation. CIGRÉ's Study Committee structure was adjusted a few times, reflecting the technical evolution of that period. In 1946, an Executive Committee was set up, National Committees were organised, and a technical periodical was issued. CIGRÉ working groups and task forces collect field data and perform system studies, and their reports are used as input for the revision of existing IEC standards. Today IEC and CIGRÉ have formal technical liaisons, and many experts are members of the relevant working groups of both IEC and CIGRÉ.

## 9.4 *THE SHORT-CIRCUIT TESTING LIAISON–STL*

The STL was founded in 1969, at a special meeting in London. The founding fathers were ASTA (Association of Short-Circuit Testing Authorities) from the United Kingdom, PEHLA (Gesellschaft für Elektrische Hochleistungsprüfungen) from Germany, KEMA (N. V. tot Keuring van Elektrotechnische Materialen) from the Netherlands, and CESI (Centro Sperimentale Italiano) from Italy. The principal purposes of STL were to establish a liaison between the four testing associations, thus ensuring an interchange of technical information and to determine if there is a need for a common certificate. In 1971, ESEF (Ensemble des Stations d'Essais à Grandes Puissance Françaises) from France joined STL and the members of STL started to formulate common interpretations of the most important IEC recommendations, commencing with IEC 56 for high-voltage switchgear. In 1975, SATS (Scandinavian Association for Testing Switchgear) from Scandinavia and the American STL/USA from North America joined as full members. Shortly thereafter the JSTC (Japanese Short-Circuit Testing Committee) from Japan became an associated member. Nowadays CPRI (Central Power Research Institute) from

India, CHPTL (China High Power Test Laboratory Liaison) from China, NEFTA (National Electric Test Facility) from South Africa, PALTS (Polish Association of Laboratories for Testing Switchgear) from Poland, VEIKI from Hungary, and ZKUSEBNICTVI from the Czech Republic participate in the technical committees from STL.

The basic aim of STL is the harmonised application of IEC and regional standards related to IEC standards to the type-testing of electrical transmission and distribution equipment (above 1000 V AC and 1200 V DC) for which type tests include high-power verification tests.

## 9.5 STANDARDS RELATED TO HIGH-VOLTAGE ELECTRICAL POWER EQUIPMENT

**IEEE Std C37.09-1999**

*IEEE standard test procedure for AC high-voltage circuit breakers rated on a symmetrical current basis.* This test procedure summarises the various tests that are made on AC high-voltage indoor and outdoor circuit breakers, except for generator circuit breakers, which are covered in IEEE Std. C37.013-1997.

**IEC 60044-1**

*IEC standard for instrument transformers.* This standard applies to newly manufactured current transformers for use with electrical measuring instruments and electrical protective devices at frequencies from 15 Hz to 100 Hz.

**IEC 60056**

*IEC standard for high-voltage alternating current circuit breakers.* This standard is applicable to AC circuit breakers designed for indoor or outdoor installation and for operation up to and including 60 Hz on systems having voltages above 1000 V. This standard does not cover circuit breakers intended for use on motive power units of electrical traction equipment; these are covered by IEC Publication 77: rules for electric traction equipment.

**IEC 60076-1**

*IEC standard for power transformers.* This part of the standard applies to three-phase and single-phase power transformers (including auto-transformers) with the exception of certain categories of small and special transformers, such as welding transformers, testing transformers, instrument transformers, transformers for static convertors, starting transformers, and traction transformers mounted on rolling stock.

### IEC 60076-5

*IEC standard for power transformers.* This part of the standard identifies the requirements for power transformers to sustain without damage from the effects of overcurrents originated by external short circuits. The requirements apply to transformers as defined in the scope of IEC 60076-1.

### IEC 60099-1

*IEC standard for surge arresters.* This part of the standard applies to surge-protective devices designed for repeated operation to limit voltage surges on AC power circuits and to interrupt power-follow current. In particular, the standard applies to surge arresters consisting of single or multiple spark gaps in series with one or more nonlinear resistors.

### IEC 60129

*IEC standard for alternating current disconnectors and earthing switches.* This standard applies to alternating current disconnectors and earthing switches designed for indoor and outdoor installation, for voltages above 1000 V and for service frequencies up to and including 60 Hz.

### IEC 60265-1

*IEC standard for high-voltage switches.* This part of the standard is applicable to three-phase alternating current switches and switch-disconnectors that have making and breaking current ratings for indoor and outdoor installations, for rated voltages above 1 kV and less than 52 kV, and for rated frequencies from 16 2/3 Hz up to and including 60 Hz.

### IEC 60282-1

*IEC standard for high-voltage current-limiting fuses.* This standard applies to all types of high-voltage current-limiting fuses designed for use outdoors or indoors on alternating current systems of 50 Hz and 60 Hz, and of rated voltages exceeding 1000 V. Some fuses are provided with fuse-links equipped with an indicating device or striker. These fuses come within the scope of this standard, but the correct opening of the striker in combination with the tripping mechanism of the switching device is outside the scope of this standard; see IEC 60420.

### IEC 60282-2

*IEC standard for expulsion fuses.* This standard specifies requirements for expulsion fuses designed for use outdoors or indoors on alternating current systems of 50 Hz and 60 Hz, and of rated voltages exceeding 1000 V.

Expulsion fuses are fuses in which the arc is extinguished by the expulsion effects of the gases produced by the arc.

### IEC 60289

*IEC standard for reactors.* This standard applies to shunt reactors, current-limiting reactors, neutral-earthing reactors, damping reactors, tuning (filter) reactors, earthing transformers (neutral couplers), and arc-suppression reactors.

### IEC 60298

*IEC standard for AC metal-enclosed switchgear.* This standard specifies requirements for factory-assembled metal-enclosed switchgear and control gear for alternating current of rated voltages above 1 kV and up to and including 52 kV for indoor and outdoor installation, and for service frequencies up to and including 60 Hz.

### IEC 60353

*IEC standard for line traps.* This standard applies to line traps inserted into high-voltage AC transmission lines to prevent undue loss of carrier signal power, typically in the range 30 kHz to 500 kHz, under all power system conditions, and to minimise interference from carrier signalling systems on adjacent transmission lines.

### IEC 60420

*IEC standard for switch-fuse combinations.* This standard applies to three-pole units for public and industrial distribution systems that are functional assemblies of switches, including switch-disconnectors and current-limiting fuses, and thus able to interrupt load currents and short-circuit currents.

### IEC 60427

*IEC standard for synthetic testing of high-voltage circuit breakers.* This standard applies to AC circuit breakers within the scope of IEC 60056. It provides the general rules for testing AC circuit breakers, for making and breaking capacities over the range of test duties described in 6.102 to 6.111 of IEC 60056, by synthetic methods.

### IEC 60517

*IEC standard for gas-insulated metal-enclosed switchgear.* This standard specifies requirements for gas-insulated metal-enclosed switchgear in which the insulation is obtained, at least partly, by an insulating gas other than air at atmospheric pressure, for alternating current of voltages of 72.5 kV and above, for indoor and outdoor installation, and for services frequencies up to and including 60 Hz.

### IEC 60694

*IEC standard with common specifications for high-voltage switchgear and controlgear standards.*

### IEC 60726

*IEC standard for dry-type power transformers.* This standard applies to dry-type power transformers (including autotransformers), having a highest voltage up to and including 36 kV.

### IEC 61128

*IEC standard for AC disconnectors.* This standard applies to alternating current disconnectors, rated 52 kV and above, capable of switching bus-transfer currents.

### IEC 61129

*IEC standard for AC earthing switches.* This standard applies to alternating current switches, rated 52 kV and above, capable of switching induced currents.

### IEC 61233

*IEC technical report for inductive load switching by high-voltage circuit breakers.* This technical report is applicable to circuit breakers that are used for switching of transformer magnetising currents, currents of high-voltage motors, or currents of shunt reactors.

### IEC 61259

*IEC standard for gas-insulated switchgear.* This standard applies to alternating current gas-insulated metal-enclosed disconnectors for rated voltages of 72.5 kV and above. It provides test requirements for gas-insulated metal-enclosed disconnectors used to switch small capacitive currents (no load currents), which occur when sections of bus bars or grading capacitors are energised or de-energised.

### IEC 61633

*IEC technical report for high-voltage metal-enclosed and dead tank circuit breakers.* This report contains information and recommendations for test circuits and procedures for type-testing relevant to short-circuit making and breaking, and switching performance of metal-enclosed and dead tank circuit breakers. Special test circuits are given and the tests described can be made, in principle, in both direct and synthetic circuits.

## 9.6 REFERENCES FOR FURTHER READING

*Inside the IEC*, IEC Central Office, Geneva, Switzerland, 1998.

*General STL Guide*, Short-circuit Testing Liaison, Asta House, Rugby, England, 1998.

Lepecki, J., "Cigré 75 years," *Électra* **167**, 5–9 (1996).

# 10 Testing of Circuit Breakers

The simulation of transient phenomena that occur in the power system is not simple. We are placed with the fact that the extensiveness and the complexity of the network has to be translated to a modest number of concentrated elements between the walls of the laboratory. The development tests for the manufacturers of switchgear, high-voltage fuses, load-break switches, and other interrupting devices, and also the acceptance tests for the utilities to verify whether the switching equipment meets the specification laid down in the standards, require accurate measurements and reproducibility of the tests.

The interrupting process in a circuit breaker, as we have seen in Chapter 4, *Circuit Breakers*, involves electromagnetic fields, thermodynamics, static and dynamic mechanics, fluid dynamics, and chemical processes. As it is difficult to give a set of formulas to design and calculate the behaviour of circuit breakers, testing is an indispensable procedure in their development and verification.

According to the IEC-60056 standard, type tests of circuit breakers include the following:

- *Mechanical and environmental tests, including mechanical operation test at ambient air temperature, low- and high-temperature tests, humidity test, test to prove operation under severe ice conditions, and static terminal load test;*
- *Short-circuit current making and breaking tests including terminal fault tests, short-line fault test, and also out-of-phase test;*
- *Capacitive current switching tests, including line-charging, cable-charging, single capacitor bank, and back-to-back capacitor bank tests; and*
- *Magnetising and small inductive current switching tests.*

Except for the mechanical and environmental tests, all the tests in the IEC-60056 type test requirements are designed to prove the interrupting ability of circuit breakers. The fault current interrupting ability is the basic function of a circuit breaker and, therefore, circuit breaker testing requires high short-circuit power in conjunction with both the current and voltage being variable over a wide range. The testing of circuit breakers can be carried out either in the actual system or in a simulated test laboratory situation. System testing offers the advantage that no special investments are necessary for the testing equipment, and the breakers face the real fault conditions as they would in service.

However, system testing is not very practical. Apart from interfering with normal system operation and security, it is difficult to create the various system conditions as prescribed by the standards. Moreover, it would not readily supply the facilities for development testing for the manufacturers. A high-power laboratory provides the possibility to test circuit breakers conveniently under the simulated system conditions.

## *10.1 THE HIGH-POWER LABORATORY*

A high-power laboratory is designed to test not only interrupting devices, such as fuses and circuit breakers, switching devices, such as load break switches and disconnectors, but also other apparatus that can face over-voltages or encounter short-circuit currents such as transformers, surge arresters, and bus bar systems conveniently under the simulated system conditions.

Based on the difference of the power supply, two types of high-power laboratories can be distinguished. One is the testing station directly supplied from the network (Figure 10.1) and the other is the specially built short-circuit generator test station (Figure 10.2).

The network testing station is built close to a high-voltage transmission substation and uses the network to provide the short-circuit power directly. To avoid instability of the supplying network during the short-circuit test, the available short-circuit power at the connection point should be approximately ten times the maximum power used during an actual test. The rather low financial investments, the relatively easy operation of the station, and the modest maintenance costs are the main advantages. For testing, the test engineers are dependent on the available power at the moment of testing, which often makes it necessary to test at night. During peak loads, for instance, in the summertime when numerous air-conditioners do their work, the test engineers are very restricted in their test program.

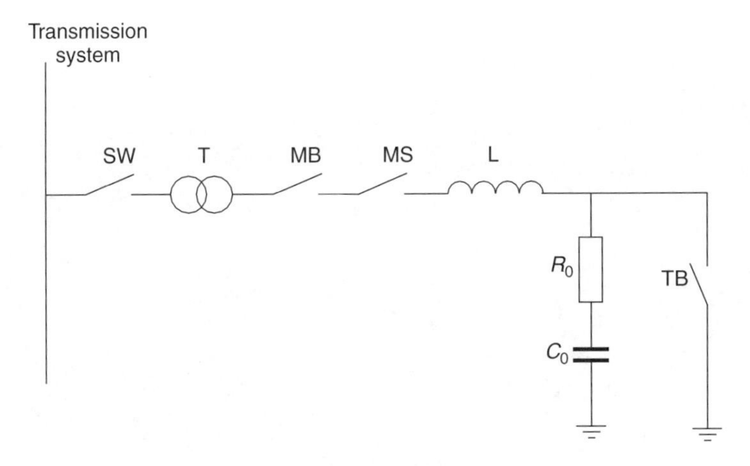


**Figure 10.1** One-line diagram of a testing station using the network as a source of supply

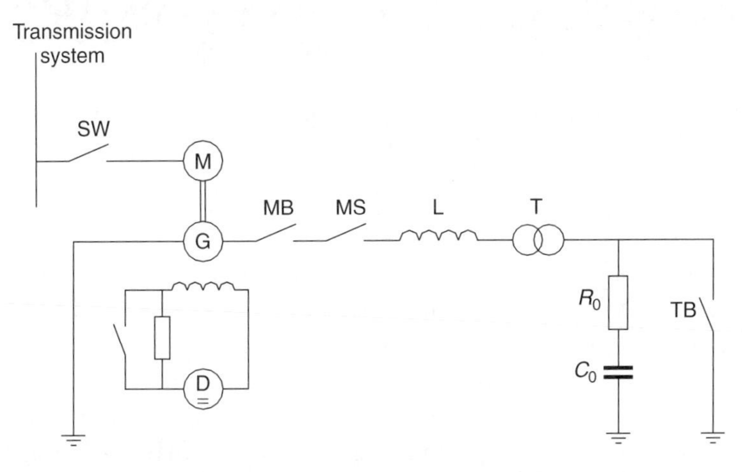


**Figure 10.2** One-line diagram of a testing station using short-circuit generators as a supply source. SW = source switch; MB = master breaker; L = current-limiting reactor; M = motor; T = short-circuit transformer; MS = make switch; $R_0$, $C_0$ = TRV adjusting elements; D = excitor; TB = test breaker; G = short-circuit generator

The generator testing station is equipped with specially designed short-circuit generators, which supply the short-circuit power. A motor is used to spin the short-circuit generator to its synchronous speed before the test takes place. During the short-circuit test, the excitation and the kinetic energy of the rotor mass, supply the power. Because it takes about twenty minutes to spin up the generator, the power taken from the network is considerably lower than the power used at the moment of testing. The largest high-power laboratory using generators, and also the largest high-power test facility in the world is at KEMA in Arnhem, The Netherlands

**Figure 10.3** KEMA's High-Power Laboratory, located on the banks of the river Rhine, in Arnhem, The Netherlands, was commissioned in 1973 and extended with a synthetic test facility in 1980 (courtesy of KEMA)

(Figure 10.3). The total available short-circuit power in the test bay at KEMA laboratories is 8400 MVA. Other large generator testing stations are located at Xian in the Peoples Republic of China, at Bangalore in India, at Bechovicze (near Prague) in the Czech Republic, and at Milano in Italy. In addition, the large switchgear manufacturers have generator testing stations: Siemens in Berlin, Germany, ABB in Ludvika, Sweden and in Baden, Switzerland, GEC-Alsthom in Villeurbanne, France, and Schneider Electric in Grenoble, France.

The largest network station in the world is operated by IREQ in Varennes in Canada. Electricité de France owns a large test facility 'les Renardieres' in Moiret-sur-Loing (near Paris) in France. Les Renardieres is also equipped with generators and can also use the network for the supply.

## *10.2 THE HISTORICAL DEVELOPMENT OF CIRCUIT BREAKER TESTING*

From the beginning, circuit breakers have been an indispensable part of power systems because they act as protective devices to switch off short-circuit currents and are used to isolate faulted sections of the network. The first circuit-interrupting devices were developed in the early years of the twentieth century by means of a trial-and-error engineering approach. Until today, circuit breaker development relies heavily on testing. The increase in the breaking capability of interrupting chambers has been closely linked with the increase in the available short-circuit power in the testing stations and the improvements made in measuring equipment to accurately record the phenomena around current zero, which has led to a better understanding of the complicated physical processes between the arcing contact during the interrupting process.

The first serious work on the study and the design of test circuits was carried out by J. Biermanns at AEG in Germany. In 1925, he proposed to test one pole of a circuit breaker at one and a half times the phase voltage by means of a so-called *Kunstschaltung*. In 1931, he improved his test circuit and used a transformer to generate a recovery voltage higher than the voltage of the current supply. In that same year, E. Marx proposed an impulse generator to stress the breaker contacts with a recovery voltage after the short-circuit current was cleared. It was in the early fifties that the interrupting capabilities of the circuit breaker began to surpass the power available from direct tests. A direct test implies a test circuit with a single power source, having an MVA rating that is at least equal to the full demand of the test. In a direct test circuit, both the interrupting current and the recovery voltage are supplied by the same power source, as can be seen from Figures 10.1 and 10.2, and the tests can be carried out three phase.

A means to overcome the shortage of test power was, for many years, the unit testing method. In this method, units, consisting of one or more interrupting chambers, are tested separately at a fraction of the rated voltage of a complete breaker pole. This technique, however, retains some problems such as the thorough understanding of the influence of the post-arc conductivity on the voltage distribution along the units. Modern $SF_6$ puffer circuit breakers have a high interrupting rating per break and this interrupting rating surpasses the short-circuit power of the largest high-power laboratories. The synthetic testing method was being developed during the 1930s to overcome the problem of insufficient direct test power. The synthetic testing method applies two source circuits to supply the high current and the recovery voltage separately. In 1942, F. Weil introduced a synthetic test circuit, which was later improved by G. Dobke and others and is now used worldwide, as the parallel current-injection circuit often referred to as the Weil-Dobke test circuit.

## *10.3 DIRECT TEST CIRCUITS*

Direct test circuits can be either fed by specially designed short-circuit generators or supplied from the network. A one-line diagram showing the principal layout of a high-power laboratory, equipped with generators, is depicted in Figure 10.2. Directly behind the specially designed *three-phase short-circuit generator* (Figure 10.4), the *master breaker* is placed (Figure 10.5). This master breaker has the duty to clear the short-circuit current in the case of a failure of the test object. After the master breaker, it is the *make switch* (Figure 10.6), which makes the short-circuit current to flow when it is closed. The *current-limiting reactor* (Figure 10.7) is used to add extra reactance in the circuit (if required) to match the current with the driving voltage. Because the terminal voltage of the short-circuit generators is relatively low (between 10 and 15 kV), specially designed *short-circuit transformers* (Figure 10.8) are necessary to transform the short-circuit power to a higher voltage level (at KEMA high-power laboratory till 245 kV to ground). In the case of a circuit breaker as a test object, the TRV-adjusting elements are usually connected at the high-voltage side

**Figure 10.4** Two of the four short-circuit generators with their auxiliary machines in KEMA's generator room (courtesy of KEMA)

**Figure 10.5** Single-phase master breaker. This air-blast breaker has an operating pressure of 80 bar and is capable of clearing 160 $kA_{rms}$ in 7 milliseconds at 15 kV. Each generator phase has its own master breaker (courtesy of KEMA)

of the transformers. At the end of the one-line diagram, we find the test object that is usually solidly grounded. The test-object can not only be a switching element, such as a high-voltage circuit breaker, a load break switch, or a disconnector, but also a bus bar, a high-voltage fuse, a surge arrester, or a transformer. The high-power laboratory contains the hardware to simulate the electromagnetic switching transients as they occur in the distributed environment of the real network.

Apart from the hardware, adequate measurement equipment and sophisticated measuring techniques are essential while performing tests in the high-power laboratory. As mentioned earlier, improvements in test techniques and measurements make higher interrupting ratings and the application of new extinguishing media possible. For any interruption test, a number of records of current and voltage are required and they must be recorded at different timescales. Other time-varying parameters, such as the position of the breaker contacts, the pressure inside the interrupting

**Figure 10.6** Single-phase make switch. Each generator phase has its own make switch, capable of making 270 $kA_{peak}$ with an accuracy of $3^{\circ}$. The maximum short-time current is 100 $kA_{rms}$ for 3 s (courtesy of KEMA)

chamber, and the temperature of the extinguishing medium describe the environment of the current-breaking process. For many years, magnetic oscillographs have been used, providing a rather slow but nevertheless rather complete record over the complete duration of a test, usually lasting a few cycles of current and voltage. With a time resolution of about 0.2 ms, the magnetic oscillograph gives an overall record of current, voltage, and the position of mechanical parts for adjusting the timing and the contact travel of the breaker contacts. The rather high-voltage insulation level of the current and voltage channels on the magnetic oscillograph has always been one of its strong advantages. The application of computer data acquisition and processing systems has resulted in the position of the traditional analogue magnetic oscillograph being replaced by the digital transient recorders in recent years. For very fast recordings, specially designed digital transient recorders are now available, which can sample every 25 ns with a resolution of 12 bits. In spite of the fact that

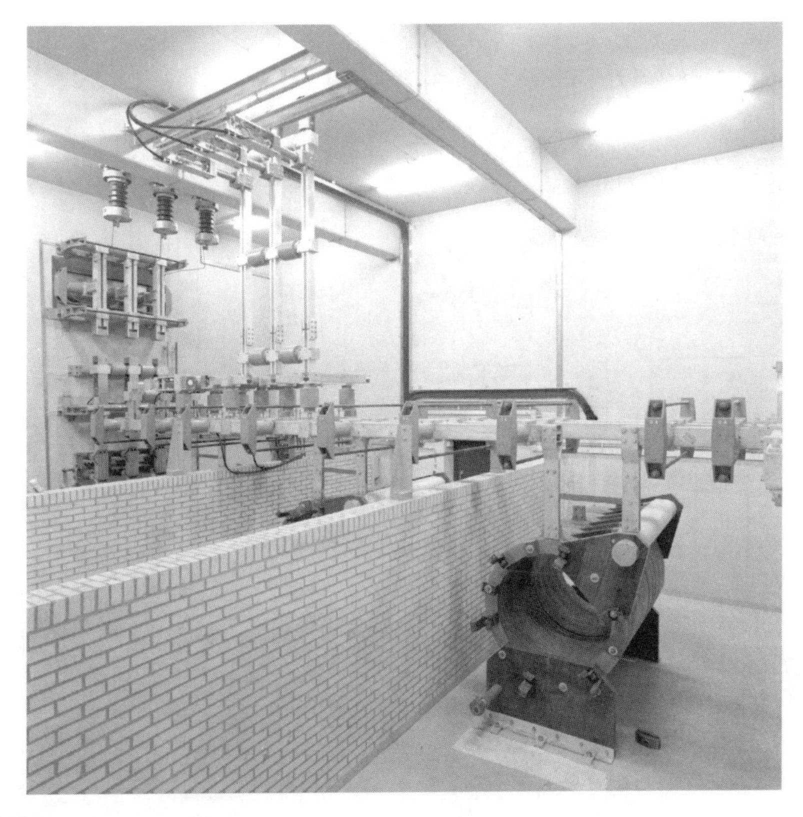

**Figure 10.7** Current limiting reactors for one short-circuit generator. (courtesy of KEMA)

the very sensitive electronic circuits make the measurement setup prone to electromagnetic disturbances, the great advantage is that once the data are stored in digital form they can be analysed by the computer and stored permanently for future study.

Although the overall accuracy of a measurement system cannot be better than the weakest link, which is usually the transducer, recent improvements in the instrumentation have made it possible to overcome many traditional problems. Optical isolators and fiber-optic transmission make it less cumbersome to eliminate electromagnetic interference in the rather unfriendly electromagnetic measurement environment of the test bay in the high-power laboratory.

In the test circuit depicted in Figure 10.9, the test breaker TB is the high-voltage breaker under test. While performing a break test, TB is in a closed position. In addition, the master breaker MB is closed, but the make switch MS is open. After the generator is spun up to the nominal power frequency and the rotor is excited to a voltage, giving the required rated testing voltage at the high-voltage side of the short-circuit transformer,

Figure 10.8 Short-circuit transformers. (courtesy of KEMA)

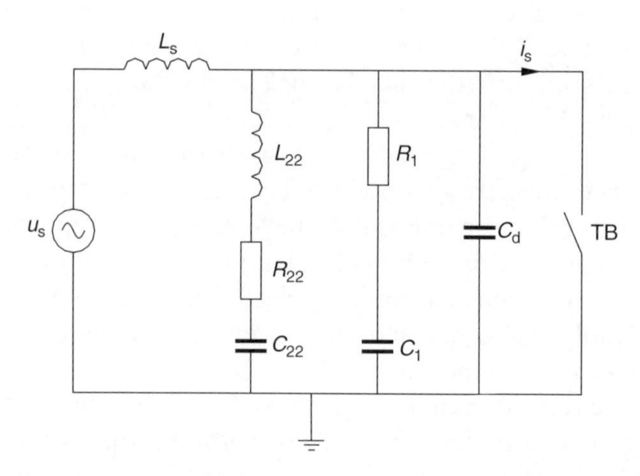


Figure 10.9 Single-phase test circuit for a short-circuit test on a high-voltage circuit breaker

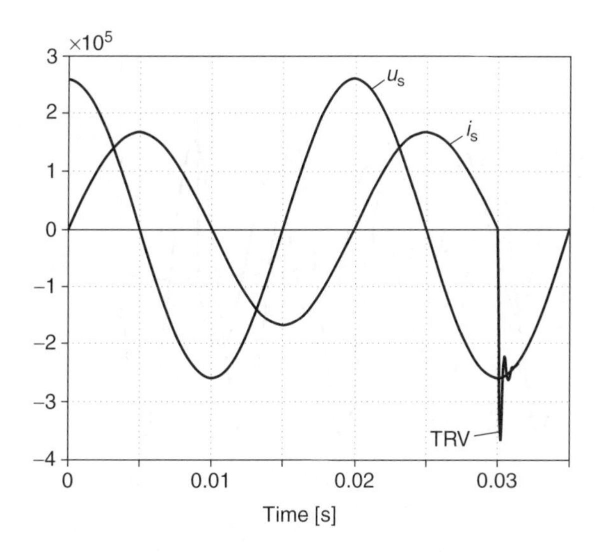


**Figure 10.10** Current and voltage traces of a single-phase current interruption

the make switch is closed and the short-circuit current flows through the TB. When the mechanism of TB receives an opening command, the breaker contacts move apart and the TB interrupts the current. After a successful interruption of the short-circuit current, the TB is stressed by the transient recovery voltage, coming from the oscillation of the TRV-adjusting elements $R_1$, $C_1$, $C_d$, $R_{22}$, $L_{22}$, and $C_{22}$ together with the inductance formed by the current-limiting reactor $L_S$, the synchronous reactance of the generator, and the leakage reactance of the short-circuit transformers. When the transient recovery has damped out, the TB faces the power frequency–recovery voltage. In Figure 10.10, the current and voltage traces of a single-phase current interruption are shown.

When the circuit breaker has to perform a make–break test, the breaker must close in on a short circuit. Before a make–break test, the TB is in open position and it closes after the MS has closed. Therefore, TB senses the power frequency supply voltage, and when the contacts of TB close the short circuit, current flows through the test circuit. In Figure 10.11, the current and voltage traces of a single-phase make–break test are shown.

The DC component of the current is determined by the instant of closing of the TB. The supply circuit is mainly an inductive circuit and this implies that if TB closes when the supply voltage is at its maximum, the DC component is zero and the current is called *symmetrical*. When TB closes at voltage zero, the current starts with maximum offset and is called *asymmetrical*. Because of the ohmic losses in the test circuit, the DC

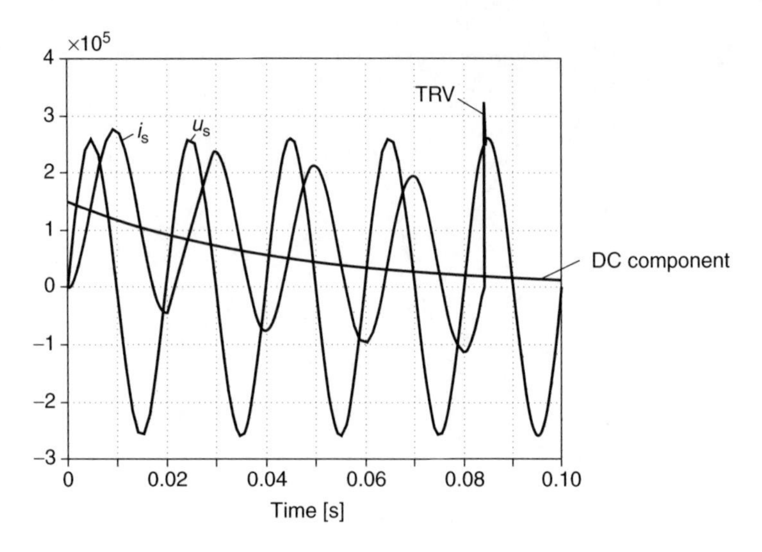


**Figure 10.11** Current and voltage traces of a single-phase make-break test

component damps out with a certain time constant. The time constant of the test circuit can be adjusted by inserting an additional resistance.

## *10.4 SYNTHETIC TEST CIRCUITS*

The increase in the interrupting capability of high-voltage circuit breakers makes it necessary to invest large amounts of money in high-power laboratories. To increase the power of a high-power laboratory by simply increasing the number of generators and transformers is neither an economical solution nor a very practical approach because the increase of the short-circuit rating of circuit breakers has gone to fast since the introduction of $SF_6$ as extinguishing medium.

When we analyse the fault-interrupting process of a circuit breaker in the actual network, two distinct intervals can be distinguished, (see Figure 10.12).

The circuit breaker is in a closed position when the fault occurs. The short-circuit is detected by a current transformer, the protection delivers a tripping signal to the breaker, and the mechanism of the breaker moves the breaker contacts apart. When the contacts part, an arc is initiated between the arcing contacts, and the interrupting chamber is designed such that at the first or second current zero crossing, the short-circuit current is interrupted and the breaker is stressed by the transient recovery voltage and the power frequency–recovery voltage. Synthetic testing methods are

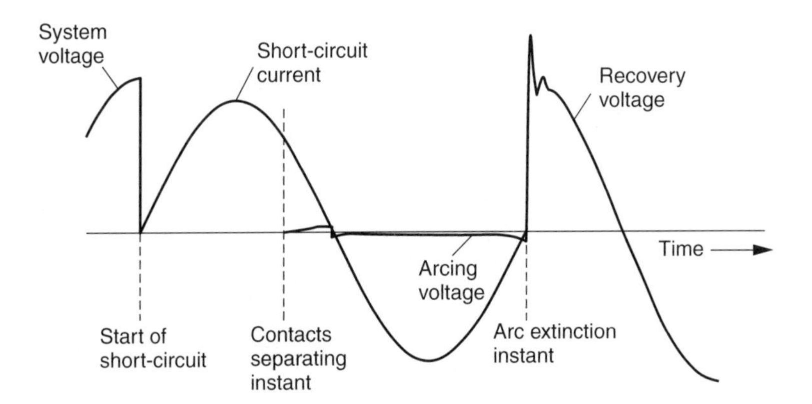


**Figure 10.12** Short-circuit current, arcing voltage, and recovery voltage for a circuit breaker clearing a system fault

based on the fact that during the interrupting process, the circuit breaker is stressed by high current and by high voltage at different time periods. This gives the possibility of using two separate sources of energy: one source supplying the short-circuit current during the arcing period and another source supplying the transient recovery voltage and the power frequency–recovery voltage. The overlap of the current and voltage source takes place during the so-called *interaction interval* around the current zero, where the current interruption takes place. Two different synthetic test methods, current injection and voltage injection, are employed as synthetic testing technique.

For the current-injection method, whereby a high-frequency current is injected into the arc of the test breaker before the short-circuit has its zero crossing, there are two principal possibilities, namely, *parallel* and *series* current injection. In addition, for the voltage injection method, whereby a high-frequency voltage is injected across the contacts of the test breaker after the short-circuit current has been interrupted, a *parallel* and a *series* scheme are possible. For the ultra-high voltage levels, when the breaker ratings exceed the 765-kV level, a combination of the current-injection and the voltage-injection circuit is used, the so-called three-circuit arrangement. The parallel current-injection circuit and the series voltage-injection test circuit are, at present, widely accepted for the testing of high-voltage circuit breakers.

The parallel current-injection method was developed by AEG in Germany. The synthetic test circuit used for the parallel current-injection test (see Figure 10.13) is often referred to as the *Weil-Dobke circuit.*

Figure 10.14 illustrates the interrupting current and recovery voltage of the test breaker in the parallel current-injection test circuit. The

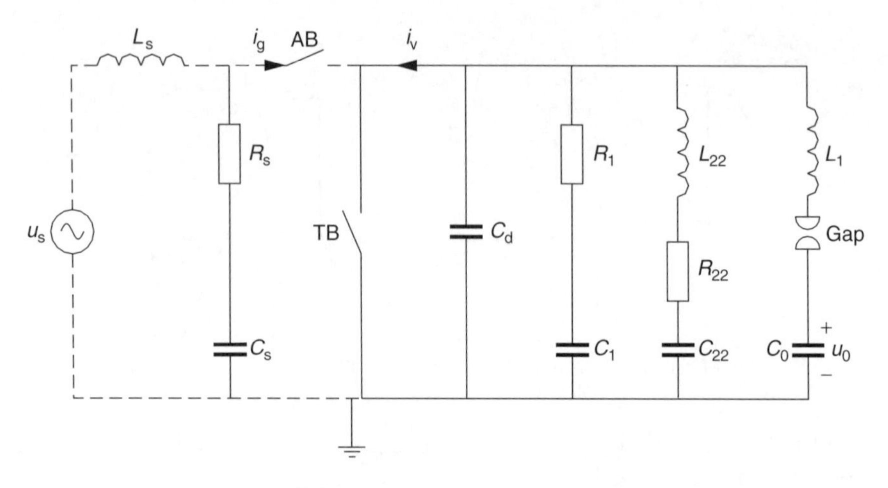


**Figure 10.13** Parallel current-injection circuit or Weil-Dobke circuit

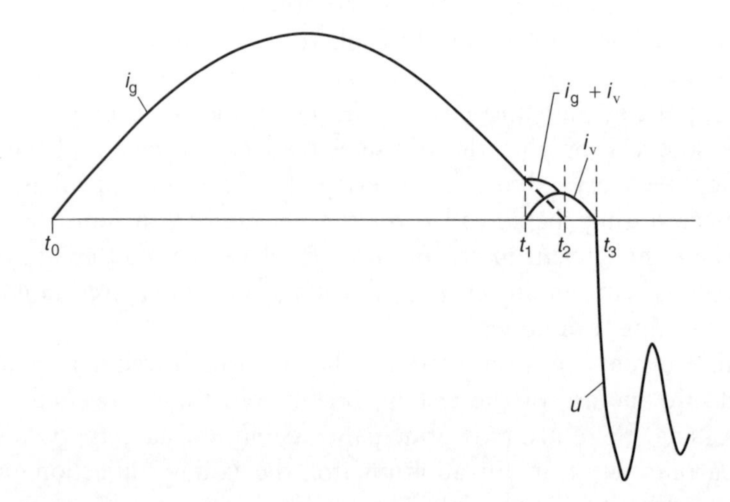


**Figure 10.14** Current and recovery voltage in the parallel current-injection scheme

short-circuit current is supplied by the short-circuit generators, which are used as the current source. Before the test, both the auxiliary breaker AB and the test breaker TB are in a closed position. At $t_0$, AB and TB open their contacts, and the short-circuit current flows through the arc channels of AB and TB. At instant $t_1$, the spark gap (see Figure 10.18) is fired and the main capacitor bank $C_0$, which was charged before the beginning of the test, discharges through the inductance $L_1$ (see Figure 10.17) and injects a high-frequency current $i_v$ in the arc channel of the TB, which adds up to the power frequency current $i_g$. During the time span from $t_1$ to $t_2$, the current source circuit and the injection circuit are connected in parallel to the TB, and the current through the TB is the sum of the short-circuit current supplied by the generator and the injected current.

At time instant $t_2$, the power frequency short-circuit current $i_g$ supplied by the generator, reaches current zero. Because the driving voltage of the generator is rather low, it is relatively easy for the AB to interrupt $i_g$, and the AB separates the high-current circuit from the high-voltage circuit. When the TB interrupts the injected current at time instant $t_3$, it is stressed by the transient recovery voltage resulting from the voltage, oscillating in the voltage circuit across the test breaker. The spark gap is still conducting during this time period.

The shape of the current waveform in the interaction interval around current zero is important because the physical processes in the arc column have a very short time constant, in the order of a few microseconds. It is, therefore, necessary to keep the rate of change $di_v/dt$, of the interrupted current just before current zero, the same as the $di/dt$ in the direct test circuit. In the time period around current zero, there is a very strong arc–circuit interaction and this time period is very critical for the circuit breaker; the current is approaching zero and at the same time the arc voltage is changing. During the high current interval, that is, before current zero when the circuit breaker is in its arcing phase, it is necessary that the energy released in the interrupting chamber is the same as the energy that would be released in a direct test, so that at the beginning of the interval of interaction when the actual current interruption takes place, the conditions in the breaker are equal to those in a direct test. The arcing energy is determined by the current and arc voltage, and care has to be given to ensure the correct magnitude and shape of the current. The magnitude of the current is controlled by the reactance of the current circuit and the driving voltage of the generator and is influenced by the arc voltage of the TB and the arc voltage of the AB. According to the IEC-60427 standard for synthetic testing of high-voltage circuit breakers, the influence of the arc voltage in a synthetic test circuit is limited to such a condition that the distorted current peak and the duration of the last current loop shall not be less than 90 percent of the values specified. This corresponds with a reduction of approximately 12 percent of the energy input in the last current loop. This leads to the requirement that for a more or less constant value of the arc voltage, the driving voltage of the supplying current source must be at least 12 times the sum of the arc voltages of the TB and the AB.

Apart from the correct rate of change $di/dt$ of the injected current (which is determined by the charging voltage $u_0$ of the main capacitor bank $C_0$ divided by the main inductance $L_1$), the frequency of the injected current is of importance. To prevent undue influence on the waveform, and subsequently on the arc energy input, the lower limit of the frequency of the injected current is fixed by the IEC-60427 standard at 250 Hz. In addition, the interval in which the arc voltage still changes significantly must be longer than the short time period $t_2$ to $t_3$ when the injected current flows through the arc channel, and therefore the maximum frequency of the injected current is fixed by the IEC-60427 standard at 1000 Hz.

Comparability tests and testing experience over the years have proven that the parallel current-injection synthetic test scheme gives the best equivalence with the direct test circuit, and, therefore, the Weil-Dobke test circuit is used by many high-power laboratories in the world. However, the Weil-Dobke circuit has some inherent limitations. During the period of current injection, the voltage on the main capacitor bank $C_0$ reverses in polarity. After interruption of the injected current by the breaker under test, the injection circuit produces the transient recovery voltage. By the time the TRV oscillations have damped out, the remaining recovery voltage is a DC voltage, and this puts a higher stress on the breaker than the power frequency AC–recovery voltage from the direct test circuit.

The difference between voltage injection synthetic test circuits and current-injection synthetic test circuits is that in the voltage-injection test circuits, the AB and the TB interrupt the short-circuit current simultaneously, and both the current source circuit and the voltage source circuit contribute to it.

Figure 10.15 shows the series voltage-injection synthetic test circuit, and the current and the recovery voltage of this test circuit are depicted in Figure 10.16.

The initial part of the TRV comes from the current source circuit and, after current zero, the voltage source circuit is injected (by firing the spark gap), supplying the rest of the TRV.

Before the test, both the AB and the TB are in a closed position. The generator in the current source circuit supplies the short-circuit current and the post-arc current for the TB and, during the whole interruption interval when the arc voltage changes significantly, the arcs in the TB and in the AB are connected in series. The AB and TB clear almost simultaneously at instant $t_1$ (see Figure 10.16). There can be a few microseconds delay for TB to clear. This depends on the value of the parallel capacitor across the AB. When TB and AB have interrupted the short-circuit current, the recovery voltage of the current source is brought to TB via the parallel capacitor. Just before this recovery voltage reaches its crest value (instant $t_2$), the spark gap is fired, and the voltage oscillation of the injection circuit

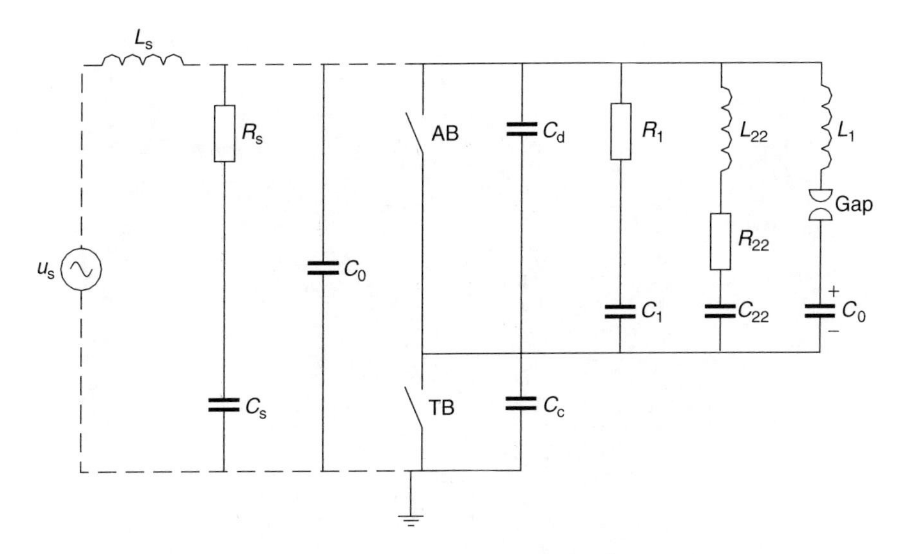


**Figure 10.15** The series voltage injection synthetic test circuit

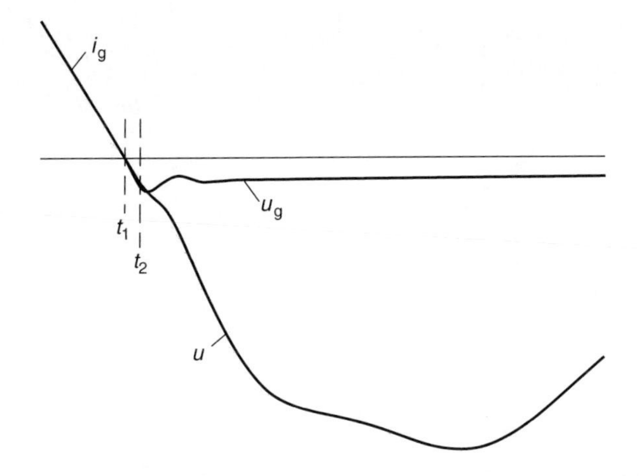


**Figure 10.16** The current and recovery voltage of the series voltage injection test circuit

is added to the recovery voltage of the current source. The resulting TRV is supplied by the voltage circuit only. Because the current injected by the voltage circuit is small, the amount of capacitive energy in the voltage circuit is relatively low. Because of this, the voltage injection method can use smaller capacitor values than the current-injection method, and is therefore more economical to use. During the interaction interval, however, the circuit parameters of the voltage-injection circuit differ from the circuit parameters of the direct test circuit. This is not the case for the current-injection circuit, and it is for this reason that the

**Figure 10.17** Main capacitor bank $C_0$ and main inductance $L_1$ of KEMA's synthetic test facility (courtesy of KEMA)

current-injection method is used when the direct test circuit reaches its testing limits.

## *10.5 SHORT-LINE FAULT TESTING*

When a fault occurs on an overhead transmission line, a few hundred meters to a few kilometers from the terminals of a high-voltage circuit breaker, we call this a *short-line fault* (Chapter 5, *Switching Transients*). Interrupting a short-line fault puts a high thermal stress on the arc-channel in the first few microseconds after current interruption because the electromagnetic waves reflecting from the short-circuit back to the terminals of the circuit breaker result in a TRV with a rate of rise from 5 to 10 kV/μs. The value of the rate of rise of the TRV at the line side depends on the interrupted short-circuit current and the characteristic impedance of the overhead transmission line.

Figure 10.18 The spark gap of KEMA's synthetic test facility can be triggered with an accuracy of microseconds (courtesy of KEMA)

In the high-power laboratory, it is not very practical to use a real overhead transmission line for short-line fault testing, and for this reason the saw-tooth shape of the line-side TRV waveform is reproduced with the help of an artificial line. An artificial line is a lumped-element network, generating the line-side TRV as specified by the IEC and ANSI standards (see Figure 10.21). The IEC 60056 standard is applicable to circuit breakers designed for direct connection to overhead transmission lines and having a rating of 52 kV and above and for a rated short-circuit breaking current exceeding 12.5 kA rms. The peak factor of the line oscillation is set at $k = 1.6$. When the specific initial transient recovery voltage (ITRV, this is the very beginning of the TRV which is mainly determined by the bus bar and line bay configuration) is not fulfilled, a line-side oscillation without time delay has to be produced by the artificial line. In the past, several types of artificial lines have been designed and built, based on series connection of LC parallel networks, a pi-network, or a T-network. These line configurations are derived from the Fourier analysis of the

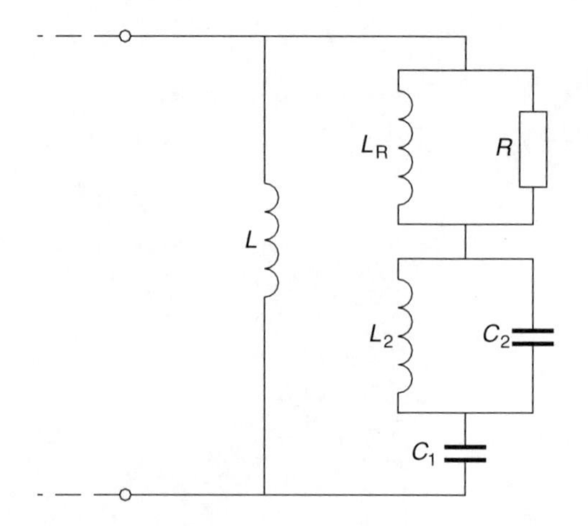


**Figure 10.19** The circuit of the KEMA artificial line

triangular waveshape of the line-side oscillation. The disadvantage of artificial lines, which are built with pi-networks, T-networks, or series LC networks is that, in theory, the number of sections must be infinite to obtain an ideal triangular-shaped waveform. A line built with a limited number of pi-sections or LC sections always has an inherent time delay caused by the capacitor seen from the terminals of the artificial line. In 1978, W. A. van der Linden from KEMA designed an artificial line for the testing of high-voltage circuit breakers up to 100 kA rms. Presently, the KEMA artificial line is used in high-power laboratories all over the world in direct and synthetic test circuits for the testing of high-voltage circuit breakers under short-line fault conditions. Figure 10.19 shows the circuit of the KEMA artificial line.

$L_R = 0.145L$, $L_2 = 0.0725L$, $C_1 = 0.333L/(Z_{line})^2$, $C_2 = 2C_1$ and $R = Z_{line}$

The main inductance $L$ reduces the current to the breakers rated short-circuit breaking current. The value of the main inductance $L$ can be calculated with

$$L = \left(\frac{1-x}{x}\right)\frac{u}{2\pi fi} \tag{10.1}$$

In this expression,

$f$ = power frequency
$u$ = phase voltage
$i$ = rated short-circuit breaking current
$x$ = the ratio between the short-line fault current and the nominal rated terminal fault current.

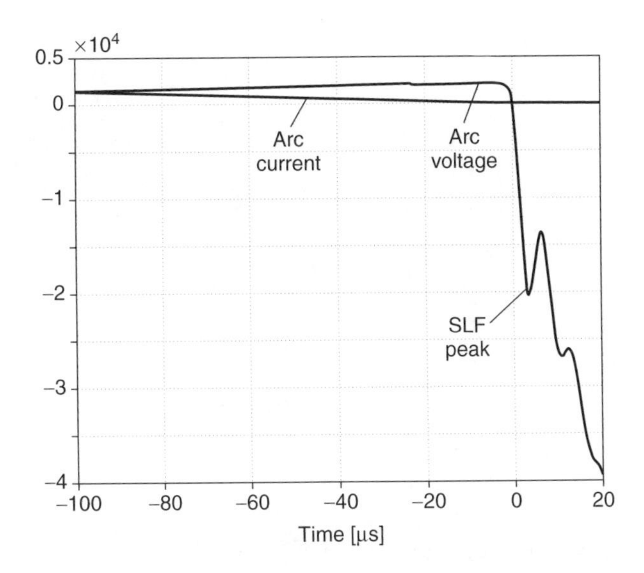


**Figure 10.20** Transient recovery voltage during the interruption of a 90 percent short-line fault of a 145 kV/31.5 kA/60 Hz $SF_6$ puffer circuit breaker

The resistor *R*, equal to the characteristic impedance of the overhead transmission line, determines the rate of rise of the line-side TRV at current zero. The paper by Van der Linden and Van der Sluis (see references) gives more insight into the design considerations of the KEMA artificial line. Figure 10.20 shows the triangular-shaped TRV, generated by the artificial line after the interruption of a 90 percent short-line fault test on a 145 kV/31.5 kA/60 Hz $SF_6$ puffer circuit breaker. The interrupted current is 28.35 kA and the rate of rise of the first excursion of the TRV is 6 kV/μs.

## *10.6 MEASURING TRANSIENT CURRENTS AND VOLTAGES*

Measuring plays an important role in the practice of electrical engineering. In fact, our knowledge and understanding of electromagnetic phenomena is based on experiments and accurate measurements. The basic physical laws have their roots in the observations made by Oerstedt, Ampere, Faraday, and Maxwell in the years between 1820 and 1870. In his '*Treatise on Electricity and Magnetism*,' Maxwell brought the existing equations together with his own concept of *displacement current*, which allowed for the continuity of current flow in nonconducting media also, in a consistent set of equations. The Maxwell equations and the wave equations for electric and magnetic fields that can be derived from them

**Figure 10.21** Artificial line to be used for short-line fault testing in a parallel current-injection circuit (courtesy of KEMA)

give a complete description of the electromagnetic phenomena and give the opportunity for a mathematical approach in electrical engineering.

In the process of developing new dielectric materials, and for the design and testing of new equipment, the predictive value of mathematics is limited, and researchers and design engineers have to rely on accurate and often sensitive measurements for the understanding of the physical processes that play a role. Measuring is not easy, and especially in an environment with strong electric fields, such as the high-voltage laboratory, or in an environment with strong electric and magnetic fields, as in the high-power laboratory, performing reliable measurements is difficult. A measurement that is carried out wrong will result in a false measuring signal and subsequently lead to a misinterpretation of what really happened. The transducers that are used to convert the quantity to be measured into a low-voltage signal must have the necessary accuracy and the correct bandwidth.

Apart from the transducer (e.g. the shunt for current measurement and the voltage divider for measuring voltages), the measurement circuit as a whole needs a careful consideration with respect to grounding and shielding.

### 10.6.1 Transducers for Current Measurements

Accurate measurement of transient currents, occurring during switching sequences in high-voltage substations or during testing in the high-power laboratory can be made with coaxial shunts (see Figure 10.23). The currents to be measured range from Amperes to kiloAmperes, and the frequency of the signals has a bandwidth from DC to megaHertz. Coaxial shunts offer several advantages; they have a relatively high output voltage, a low input impedance, are unaffected by stray fields, and are capable of measuring from DC to megaHertz. The disadvantage of the coaxial shunt is that it has to be directly connected in the primary circuit and must be mounted at ground potential. A high output voltage and a high-frequency response, impose conflicting requirements on coaxial shunt design. For a high output voltage, a high resistivity of the tube is required. This means less copper, and the inductance of the resistance tube plays a more dominant role. More copper increases the bandwidth (and also the thermal and mechanical strength), and to reduce the influence of high frequency effects, such as the skin effect, thin wall resistance tubes are required. Figure 10.22 shows the basic coaxial shunt. The output terminals for the measuring signal are in field-free environment because

$$\oint H.\mathrm{d}l = 0, \tag{10.2}$$

and therefore there is no influence of the magnetic field surrounding the current to be measured on the measurement signal.

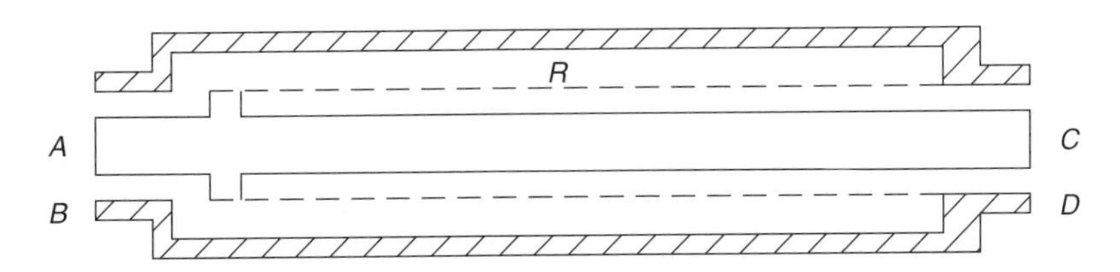


**Figure 10.22** Cross-section of a coaxial shunt. When $A$ and $B$ are used as input terminals for the current $C$ and $D$ serve as output terminals for the measuring signal and vice versa. $R$ is the calibrated resistance tube

**Figure 10.23** Coaxial shunt for measuring currents up to 100 $kA_{rms}$ with a bandwidth of 0–1 MHz (courtesy of KEMA)

The bandwidth of a shunt is determined by measuring the transient response after applying a step form current pulse to the shunt. From the transient response signal, the rise time is measured. The rise time is the time measured between 10 percent and 90 percent of the amplitude of the response signal (see Figure 10.24).

The bandwidth can easily be calculated with

$$\text{Bandwidth} = \frac{0.35}{\text{rise time } (\mu s)} \text{ MHz} \tag{10.3}$$

For current measurements at high voltage, shunts cannot be used because they have to be mounted at ground potential. Current transformers are for accurate measurements of transient currents – not really an option because of their limited bandwidth. The eddy current losses in the magnetic core increase with the frequency and limit the upper part from the measuring spectrum, whereas the resonance between the inductance of the current

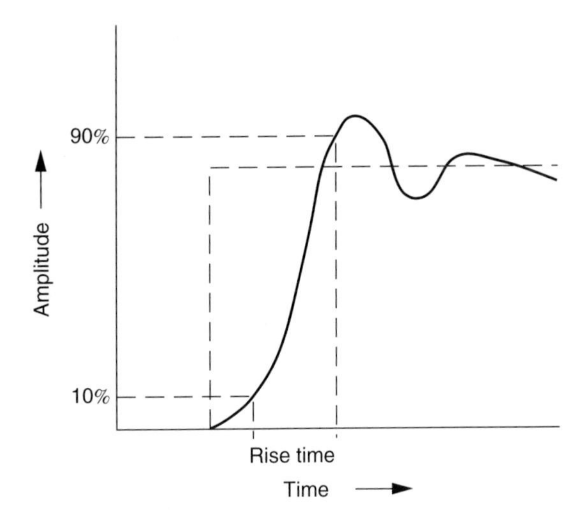


**Figure 10.24** Transient response of a shunt after applying a step form current pulse

transformer and the stray capacitance of the windings can determine the cut-off frequency. Low-frequency currents cannot be measured correctly because of saturating effect in the magnetic core. Because of core saturation, it is not possible to record the DC component in short-circuit currents. The use of large magnetic cores improves the frequency characteristic of the current transformer considerably, and time constants of twenty-five seconds can be created, but the increase in the price of such devices is more than proportional.

One of the reasons that current transformers are used for protection and measurement applications is that they are able to produce the high power output needed by electromechanical measuring equipment.

Another way of performing potential-free current measurements is by the use of Rogowski coils. A Rogowski coil (Figure 10.26) is a uniformly wound coil of constant cross-sectional area on a nonmagnetic former shaped into a closed loop around the current conductor and that senses the magnetic field around a closed path. The simplest example is an air-cored toroid. From most viewpoints, the Rogowski coil is an ideal current transducer for measuring transient currents. It gives an isolated current measurement, it does not saturate with high currents, it has an excellent bandwidth (typically from 0.1 Hz to over 1 MHz) comparable with the coaxial shunt, it does not load the circuit, and it is inexpensive to make. Only the DC component of short-circuit currents cannot be accurately measured. The principal of the Rogowski coil has been known since 1912.

The coil is placed around the conductors whose currents are to be measured as shown in Figure 10.25.

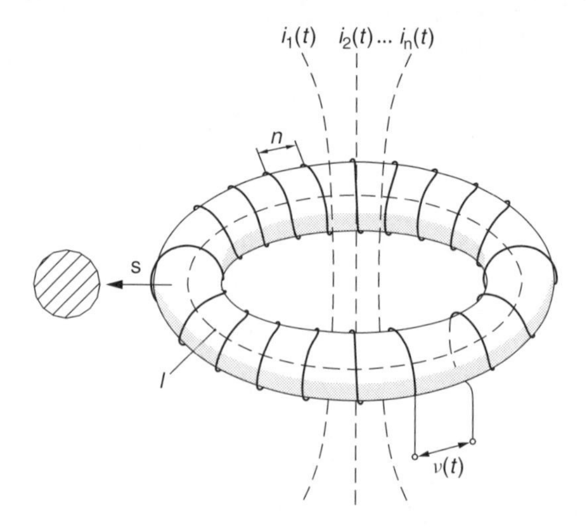


**Figure 10.25** Design principal of the Rogowski coil

**Figure 10.26** Rogowski coils for current measurements (courtesy of KEMA)

The voltage induced in the coil is

$$v(t) = -\frac{\mathrm{d}}{\mathrm{d}t}\sum_{j=1}^{N}\Phi j \tag{10.4}$$

The total number of turns of the coil is $N$, and $\Phi j$ is the instantaneous flux for the $j$th turn. The coil output voltage is $v(t)$. When $l$ is the core length, $n$ is the number of windings per unit length, $S$ is the core cross-section, and $B$ is the magnetic flux density, the total flux $\psi$ is given by

$$\psi = \oint \mathrm{d}l \iint B.n\,\mathrm{d}S \tag{10.5}$$

If the core has a constant cross-section and the wire is wound perpendicularly on the middle line $l$ with constant density, the coil output voltage is then

$$v(t) = -\mu_0 nS\frac{\mathrm{d}}{\mathrm{d}t}\left[\sum_{j=1}^{N} i_j(t)\right] = -M\frac{\mathrm{d}}{\mathrm{d}t}\left[\sum_{j=1}^{N} i_j(t)\right] \tag{10.6}$$

In this equation, $\mu_0$ is the permeability, of free space, $i(t)$ is the primary current to be measured, and $M$ is the coils mutual inductance. The output voltage of the Rogowski coil is proportional to the rate of change of the measured current. If the design considerations are followed properly, the coil's mutual inductance $M$ is independent of the conductor location inside the coil loop. To prevent the influence of magnetic fields from other current-carrying conductors in the vicinity (which is often the case in 2-phase or 3-phase systems), Rogowski coils must be designed with two-wire loops connected in the electrically opposite directions. This will cancel all electromagnetic fields from outside the coil loop and can be achieved by returning the wire through the centre of the winding, as shown in Figure 10.27.

To obtain a signal proportional to the primary current, the coil output voltage must be integrated. When considering the application of a simple RC integrator, the best results are obtained by terminating the coil with a very low resistance, thereby making the coil 'self-integrating.' Composing an RC integrator, having a time constant of 10 s by using, for example, a 1 MΩ resistor and a 10 μF capacitor, will result in a limited bandwidth because of stray capacitance of the resistor and the dielectric losses in the capacitor. Therefore, it is much better to use an active integrator,

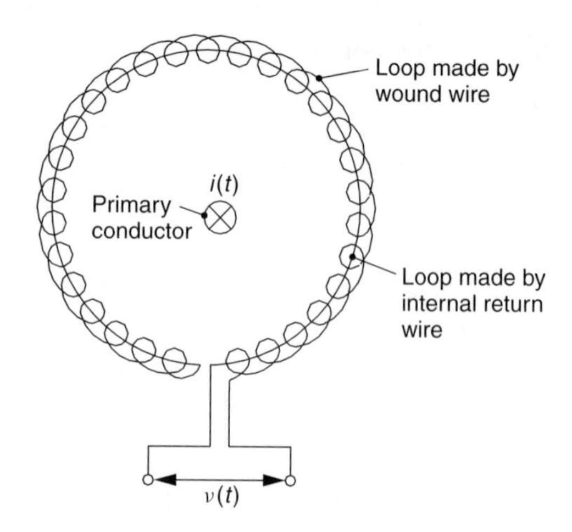


**Figure 10.27** Rogowski coil with the return wire through the centre of the winding

as described by J. A. J. Pettinga and J. Siersema in, '*A Polyphase 500 kA Current Measuring System with Rogowski Coils*'.

In 1845, Faraday found a difference in the refractive index in glass for left-handed and right-handed circularly polarised light under the influence of a magnetic field. Verdet showed in 1854 that the angle of rotation of linearly polarised light is proportional to the strength of the magnetic field and the cosine of the angle between the field and the propagation direction of the light wave. Because the limitations of this method were great, the application of the method sank into obscurity at the end of the nineteenth century, after the invention of the cathode-ray tube. In the late sixties, with the appearance of flint glass, and in the seventies, with the development of fibre sensors in combination with laser sources, the optical current transformer came within reach. Presently, optical current transformers can be bought from the major current transformer manufacturers.

## 10.6.2 Transducers for Voltage Measurements

For the measurement of high voltages at high potential, it is necessary to get a measurement signal of a considerably smaller amplitude and at ground potential; otherwise, the measurement equipment will be damaged. Because of their bad frequency characteristic, voltage transformers are not suitable for the measurement of transient voltages. A voltage divider is much more appropriate for this purpose. The divider has to be designed according to its application, given by the voltage level, the type and shape of the voltage, and sometimes by the input impedance of the connected measurement system. Therefore, it is useful to have a closer look at the different types of dividers used, by starting with a very general equivalent circuit for a divider and to derive general results for their applicability.

Voltage dividers are often built not only with resistor or capacitor elements but also in combination with a mixed resistor and capacitor type of divider. These resistor and capacitor elements are installed within an insulating cylinder with the high-voltage terminal at one end and the ground terminal at the other end. The height of this cylinder depends on the flashover voltage of the voltage applied. A rule of thumb is 5 m per megavolt for AC voltages and from 2.5 to 3 m for DC voltages. Because the active resistor and capacitor elements cannot be effectively shielded, the electrical field distribution along the divider stack can easily be disturbed by surrounding structures. This field distribution is theoretically taken into account by assuming stray capacitances to ground that will strongly influence the frequency characteristic of the voltage divider. Figure 10.28 shows the generalised equivalent circuit for a voltage divider.

The generalised equivalent circuit is a combination of $n$ equal impedances $Z_L$ and an equal number of impedances $Z_g$ to ground. The output voltage of the divider is generated by the current through the last impedance $Z_L$ to ground. The frequency characteristic of the voltage divider is dependent upon the transfer function

$$H_t(p) = n\frac{U_2}{U} = n\frac{\sinh \frac{1}{n}\sqrt{\frac{Z_L(p)}{Z_g(p)}}}{\sinh \sqrt{\frac{Z_L(p)}{Z_g(p)}}} \tag{10.7}$$

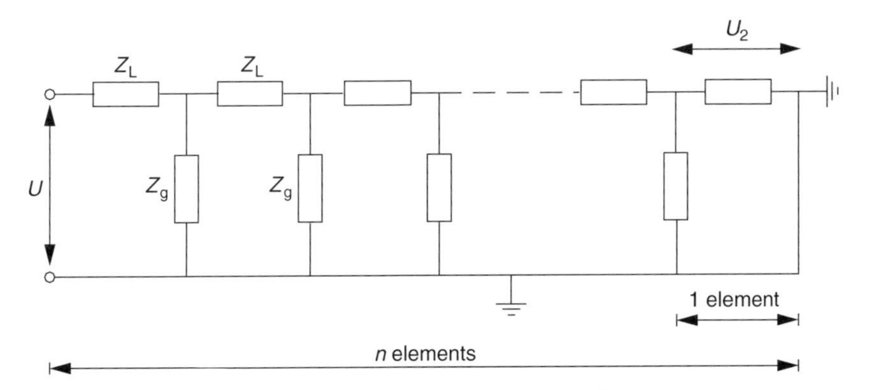


Figure 10.28 Generalised equivalent circuit for a voltage divider

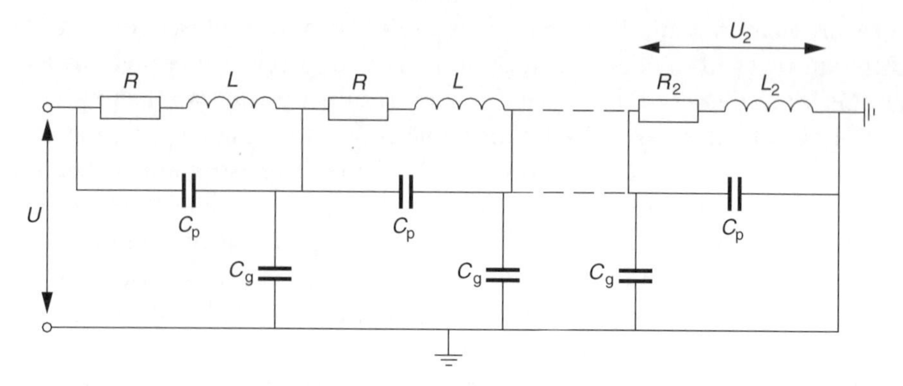


**Figure 10.29** Equivalent circuit for a resistor divider

The transfer function is, therefore, dependent upon the impedances $Z_L(p)$ and $Z_g(p)$ and can be discussed only individually, depending on the type of divider.

Resistor dividers are used for AC and DC voltage measurements, and an equivalent circuit for such a divider is shown in Figure 10.29.

The stray capacitances $C_g$ to ground, represent the effective electrical field lines to ground and $C_p$ corresponds to the ever-present stray capacitance of each resistor element. The inductive components $L$ of the resistor elements have to be taken into account only for low ohmic dividers. The normalised transfer function of a resistor divider is

$$H_t(p) = n\frac{\sinh \frac{1}{n}\sqrt{\frac{(R + pL)pC_g}{1 + (R + pL)pC_p}}}{\sinh \sqrt{\frac{(R + pL)pC_g}{1 + (R + pL)pC_p}}} \tag{10.8}$$

The applicability of these type of dividers depends strongly upon the product $RC_g$. Resistor dividers are well-suited for DC measurements and for the recording of very low frequencies ($p \to 0$). For DC measurements, the normalised transfer function will be

$$U_2 = U\frac{R_2}{R_1 + R_2} \quad R_1 = (n - 1)R \quad R_2 = R$$

The applicability of pure resistor dividers for the measurement of very high AC voltages is limited because the current flowing through the resistor elements heats up the resistors and changes the resistor values.

**Figure 10.30** Mixed divider for measuring TRVs upto 500 $kV_{peak}$ with a bandwidth of 0–1 MHz (courtesy of KEMA)

A mixed divider (Figure 10.30) with capacitive elements placed in parallel to the resistor elements combines the best of two worlds, and voltage dividers can be built with a high bandwidth that are capable of measuring DC voltages. When voltage dividers are built with pure capacitive elements, they cannot measure DC voltages but the advantage is that the loading effects on the test object are small.

## *10.7 MEASUREMENT SETUP FOR TRANSIENT VOLTAGE AND CURRENT MEASUREMENTS*

Measuring transient voltages and currents in the field or in the high-power laboratory is generally accompanied by *electromagnetic interference* (EMI). Electromagnetic fields penetrate the often imperfectly shielded recorders and induce noise directly into the amplifiers and the associated electronic circuits. Quasi-static electric and magnetic fields penetrate the braid mesh of the coaxial cables. Electric fields directly induce, via capacitive coupling, voltages on the central conductor and cable shield currents generate additional noise via the coupling impedance or transfer impedance of a coaxial cable.

For safety reasons, instrument cabinets are usually grounded and so are the coaxial shunt and the voltage divider. If we also realise that the 230-volt AC main supply has a connection with the grounding of the building and with the grounding of the distribution transformer of the local utility, we face the problem of multiple grounds. Multiple grounds introduce ground loops in which electromagnetic fields can induce voltages that not only affect the original measurement signals, but also may cause potential differences between different ground terminals inside the laboratory.

A typical measurement setup for a combined current–voltage measurement (for instance, the recording of the interruption of a short-line fault) is depicted in Figure 10.31.

The coaxial shunt is grounded and so is the low-voltage part of the voltage divider. As signal cables, single-screened 50-Ω coaxial cables are applied. The recording instrument, which can be a classical cathode ray oscilloscope or a sensitive transient recorder, can be touched during test by the technicians and is grounded for the same reason. To avoid a connection with the grounding of the main supply, an isolating transformer is placed between the recording instruments and the 230-volt AC supply. An

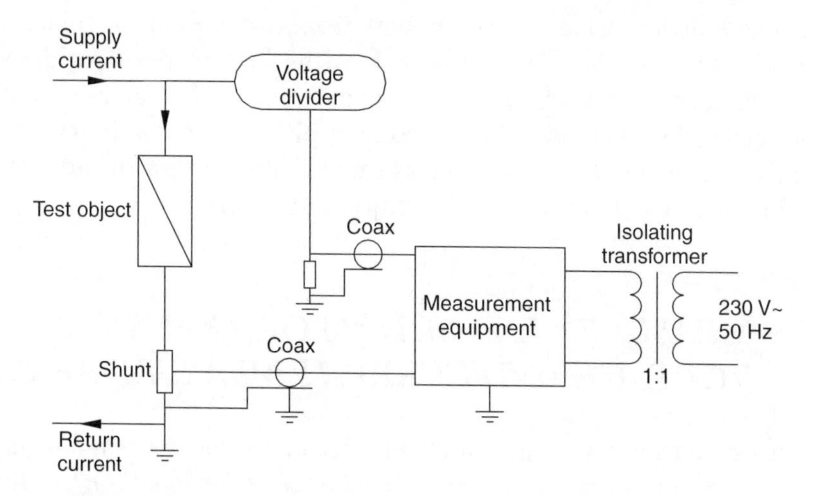


**Figure 10.31** Measurement setup for a combined current–voltage measurement

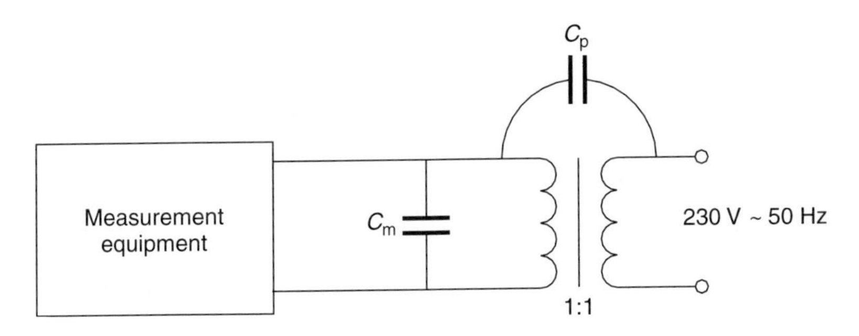


**Figure 10.32** Supply of 230 volts to recording equipment via an isolating transformer

isolating transformer is an essential piece of equipment while performing sensitive measurements in a substation, in a high-voltage laboratory, or in a high-power laboratory. Such an isolating transformer is a 1:1 transformer with a screen between the primary and secondary windings (Figure 10.32). This screen reduces the stray capacitance $C_p$ between the input and output terminals until not more than a few picoFarad. The capacitance $C_m$ seen at the 230-volt terminals of the recording equipment is much bigger. When electrical disturbances are present on the 230-volt mains supply, they will divide over $C_p$ and $C_m$ and the major part of the disturbances will appear over the smallest capacitance $C_p$. The smaller the $C_p$, the less electrical disturbance will come out of the 230-volt supply into our recording instruments. It is important that the screen between the primary and secondary winding is not grounded because this will increase the value of the capacitance $C_p$ considerably and subsequently reduce the noise protection level of the isolating transformer. In addition, the 230-volt safety ground (if present) may not be connected because an unwanted ground loop is introduced. For the technician operating the recording equipment, the use of an isolating transformer creates an unsafe situation; therefore, during the actual test, the measuring equipment should not be touched.

The measurement setup as shown in Figure 10.31 is far from ideal and can be improved on several points. Firstly, we should take care that there is only one point grounded in the circuit. If we consider the voltage and current transducers, then it is clear that because the coaxial shunt must always be grounded, the voltage divider must float. For a voltage measurement the low-voltage terminal of the divider is connected with the grounding terminal of the coaxial shunt via the screens of the coaxial cables and the metal housing of the recording equipment. The recording equipment itself is ungrounded and floats from earth (see Figure 10.33).

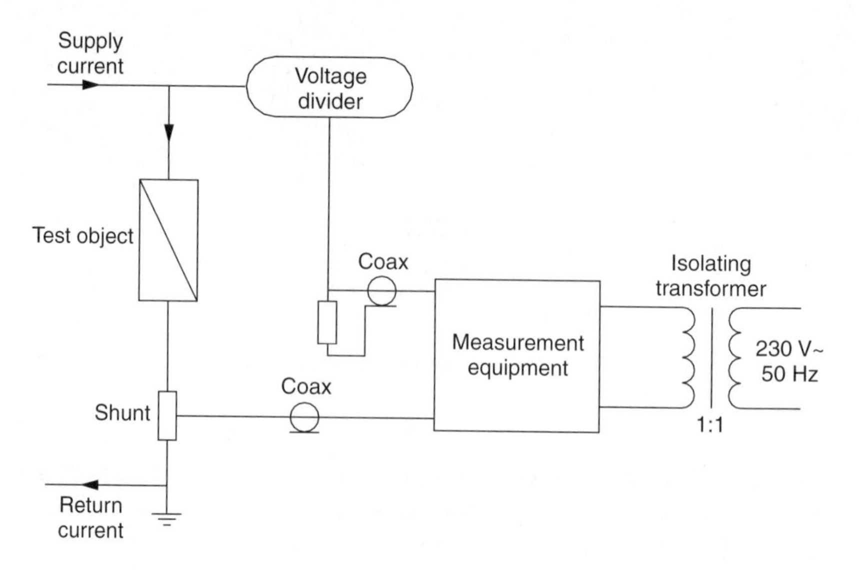


**Figure 10.33** Improved measurement setup for a combined current–voltage measurement

In the improved measurement setup, as depicted in Figure 10.33, ground loops are avoided and so an important source of disturbances is eliminated. If further improvement is necessary, coaxial cables with a lower transfer impedance or twin coaxial cables can be used or the measurement cables can be laid in steel conduit pipes, which have to be grounded at both ends, thus creating an extra conducting path for currents induced by electromagnetic fields that couple in from the environment. In this way, the transfer impedance of the signal cables is reduced and the steel conduit pipes introduce additional screening. Another measure that can be taken is to put the recording equipment in an EMC cabinet that serves as an extra metal housing around the recorders.

The shields of the coaxial cables (or the conduit pipes when the signal cables are laid in such pipes) are connected with the EMC cabinet and the mains supply is brought in via a filter. Such an EMC cabinet can be grounded for safety reasons without causing any problem for the measurement. Figure 10.34 shows the layout of a simple but effective EMC cabinet. The backside of such an EMC cabinet can be left open for ventilation and easy access to the technician as this hardly influences the screening effect. A clear and profound treatment of the problems with electromagnetic compatibility in high-voltage engineering can be found in '*Electromagnetic Compatibility in High-Voltage engineering*' by M. A. van Houten.

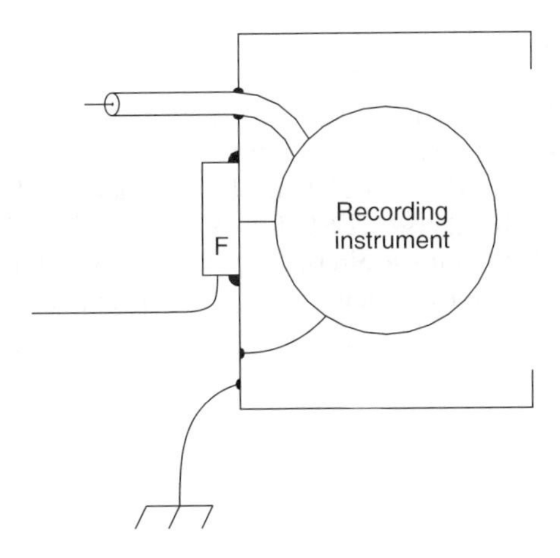


Figure 10.34 An EMC cabinet diverts the common mode currents from the shielding of the measurement cables

# 10.8 REFERENCES FOR FURTHER READING

Anderson, J. G. P., *et al.*, "Synthetic testing of AC circuit breakers, Part 1: Methods of testing and relative severity, Part 2: Requirements for circuit-breaker proving," *IEE Proceedings* **113**(4), 611–621 (1966); *IEE Proceedings* **115**(7), 996–1007 (1968).

Biermanns, J. and Hochrainer, A., "Hochspannungs-Schaltgeraete," *AEG Mitteilungen* **47** (H 7/8), S209–S212 (1957).

Blahous, L., "The problem of defining a test circuit for circuit breakers in terms of prospective voltage," *IEE Proceedings* **126**(12), 1291–1294 (1979).

Garzon, R.D., *High-Voltage Circuit Breakers: Design and Application*, Chapter 8, Marcel Dekker, 1997.

Guy St-Jean and Landry, M., "Comparison of waveshape quality of artificial lines used for short-line fault breaking tests on HV circuit breakers," *IEEE T-PWRD* **4**(4), 2109–2113 (1989).

Guy St-Jean., "Power systems transient recovery voltages," Chapter 4, IEEE Seminar/Report # 87TH0176-8-PWR, 44–58 (1987).

Kojovic, L., "Rogowski coils suit relay protection and measurement," *IEEE Computer Applications in Power* **10**(3), 47–52 (1997).

Malewski, R., "Micro-ohm shunts for precise recording of short-circuit currents," *IEEE T-PAS* **96**(2), 579–585 (1977).

Maxwell, J. C., *A Treatise of Electricity and Magnetism*, Clarendon Press, Oxford.

Pettinga, J. A. J. and Siersema, J., "A polyphase 500 kA current measuring system with Rogowski coils," *IEE-proceedings* **130**(5), Part B, 360–363 (1983).

Slamecka, E., "Systeme synthetischer Pruefschaltungen," *ETZ-A*, **84**(H18), S581–S586 (1963).

Slamecka, E., *Pruefung von Hochspannungs-Leistungsschaltern*, Part V, Chapter 3, Springer-Verlag, New York, 1966.

Slamecka, E., and Waterschek, W., *Schaltvorgaenge in Hoch- und Niederspannungsnetzen*, Chapter 5, Siemens Aktiengesellschaft, Berlin, 1972.

Slamecka E., Rieder W., and Lugton W. T., "Synthetic testing: the present state," *IEE Proc.* **125**(12), 1376–1380 (1978).

van Houten, M. A., Electromagnetic compatibility in high-voltage engineering, Thesis, Eindhoven University of Technology, (1990).

van der Linden, W. A., and van der Sluis L., "A new artificial line for testing High-Voltage Circuit Breakers," *IEEE T-PAS* **102**(4), 797–803 (1983).

"Synthetic testing of high-voltage alternating current circuit breakers," *IEC* **427**, International Electrotechnical Commission (IEC), 1989.

# Index